VOLUME FORTY EIGHT

PROFILES OF DRUG SUBSTANCES, EXCIPIENTS, AND RELATED METHODOLOGY

Contributing Editor
AHMED H. BAKHEIT

Founding Editor
KLAUS FLOREY

VOLUME FORTY EIGHT

Profiles of
DRUG SUBSTANCES, EXCIPIENTS, AND RELATED METHODOLOGY

Edited by

ABDULRAHMAN A. AL-MAJED

Department of Pharmaceutical Chemistry,
College of Pharmacy, King Saud University,
Riyadh, Kingdom of Saudi Arabia

ACADEMIC PRESS

An imprint of Elsevier

ELSEVIER

Academic Press is an imprint of Elsevier
50 Hampshire Street, 5th Floor, Cambridge, MA 02139, United States
525 B Street, Suite 1650, San Diego, CA 92101, United States
The Boulevard, Langford Lane, Kidlington, Oxford OX5 1GB, United Kingdom
125 London Wall, London, EC2Y 5AS, United Kingdom

First edition 2023

ISBN: 978-0-443-19382-8
ISSN: 1871-5125 (Series)

For information on all Academic Press publications
visit our website at https://www.elsevier.com/books-and-journals

Publisher: Zoe Kruze
Developmental Editor: Naiza Ermin Mendoza
Production Project Manager: Abdulla Sait
Cover Designer: Christian Bilbow

Typeset by STRAIVE, India
Transferred to Digital Printing 2023

Working together
to grow libraries in
developing countries

www.elsevier.com • www.bookaid.org

Contents

4. Vandetanib   **109**

Ahmed I. Al-Ghusn, Ahmed H. Bakheit, Mohamed W. Attwa, and
Haitham AlRabiah

5. Lapatinib: A comprehensive profile   **135**

Ahmed A. Abdelgalil and Hamad M. Alkahtani

6. Pharmaceutical based cosmetic serums   **167**

Nimra Khan, Sofia Ahmed, Muhammad Ali Sheraz, Zubair Anwar, and
Iqbal Ahmad

Contributors

Ahmed A. Abdelgalil
Central Laboratory, College of Pharmacy, King Saud University, Riyadh, Kingdom of Saudi Arabia

Iqbal Ahmad
Department of Pharmaceutical Chemistry, Baqai Institute of Pharmaceutical Sciences, Baqai Medical University, Karachi, Pakistan

Sofia Ahmed
Department of Pharmaceutics, Baqai Institute of Pharmaceutical Sciences, Baqai Medical University, Karachi, Pakistan

Ahmed I. Al-Ghusn
Department of Pharmaceutical Chemistry, College of Pharmacy, King Saud University, Riyadh, Kingdom of Saudi Arabia

Hamad M. Alkahtani
Department of Pharmaceutical Chemistry, College of Pharmacy, King Saud University, Riyadh, Kingdom of Saudi Arabia

Ahmed M. Alomar
Department of Pharmaceutical Chemistry, College of Pharmacy, King Saud University, Riyadh, Kingdom of Saudi Arabia

Haitham AlRabiah
Department of Pharmaceutical Chemistry, College of Pharmacy, King Saud University, Riyadh, Kingdom of Saudi Arabia

Zubair Anwar
Department of Pharmaceutical Chemistry, Baqai Institute of Pharmaceutical Sciences, Baqai Medical University, Karachi, Pakistan

Mohamed W. Attwa
Department of Pharmaceutical Chemistry, College of Pharmacy, King Saud University, Riyadh, Kingdom of Saudi Arabia

Ahmed H. Bakheit
Department of Pharmaceutical Chemistry, College of Pharmacy, King Saud University, Riyadh, Kingdom of Saudi Arabia; Department of Chemistry, Faculty of Science and Technology, Al-Neelain University, Khartoum, Sudan

Hany Darwish
Department of Pharmaceutical Chemistry, College of Pharmacy, King Saud University, Riyadh, Kingdom of Saudi Arabia; Analytical Chemistry Department, Faculty of Pharmacy, Cairo University, Cairo, Egypt

Ibrahim A. Darwish
Department of Pharmaceutical Chemistry, College of Pharmacy, King Saud University, Riyadh, Kingdom of Saudi Arabia

Nimra Khan
Department of Pharmacy Practice, Baqai Institute of Pharmaceutical Sciences, Baqai Medical University, Karachi, Pakistan

Muhammad Ali Sheraz
Department of Pharmacy Practice; Department of Pharmaceutics, Baqai Institute of Pharmaceutical Sciences, Baqai Medical University, Karachi, Pakistan

Preface

The profiling of drug substances and excipients has become an invaluable tool for the pharmaceutical analyst in his quest for the best quality analytical data. Volume 48 contains five profiles, namely Brimonidine, Crizotinib, Remdesivir, Vandetanib, and Lapatinib in addition to a chapter discussing pharmaceutical-based cosmetic serums. A profile of this series is similar to a full monograph for a drug substance or excipient, including a description of nomenclature, chemical names, proprietary and nonproprietary names, formulas and structures, appearance, and compendia information following a brief introduction. Physical parameters such as UV, IR, MS, NMR, X-ray diffraction, optical rotation, thermal techniques, hygroscopicity, dissociation constants, solubility, and partition coefficients are then discussed. The book includes a comprehensive list of analytical methods, including titrimetric, spectrophotometric, chromatographic, immunoassay, and radioactive labeling techniques. Stability, pharmacokinetics, metabolism, and excretion as well as pharmacology and toxicology are covered briefly.

The chapter writers' contributions as well as the quality of the work will continue to be critical to the success of the Profiles of Drug Substances, Excipients, and Related Methodology series. It's somewhat ironic that in today's drug development atmosphere, the requirement for these types of Profiles has become increasingly critical.

We hope that this series of books will serve as a significant resource for the pharmaceutical community, scientific and industrial researchers, pharmacists, and pharmacy students in their daily work.

Many thanks to Naiza Mendoza (Elsevier's Developmental Editor) and Abdulla Sait (Elsevier's Editorial Project Manager) for their tireless efforts in the production of this volume. Also my sincere thanks go to the authors and their coauthors for the valuable contributions in this series.

I welcome correspondence from any member of the pharmaceutical community who wishes to offer an opinion or input.

ABDULRAHMAN A. AL-MAJED

CHAPTER ONE

Brimonidine

Ahmed H. Bakheit[a,b], Ahmed M. Alomar[a], Hany Darwish[a,c], and Hamad M. Alkahtani[a]

[a]Department of Pharmaceutical Chemistry, College of Pharmacy, King Saud University, Riyadh, Kingdom of Saudi Arabia
[b]Department of Chemistry, Faculty of Science and Technology, Al-Neelain University, Khartoum, Sudan
[c]Analytical Chemistry Department, Faculty of Pharmacy, Cairo University, Cairo, Egypt

Contents

Profiles of Drug Substances, Excipients, and Related Methodology, Volume 48
ISSN 1871-5125
https://doi.org/10.1016/bs.podrm.2022.11.001

1

1. Description

1.1 Nomenclature

1.1.1 Systematic chemical names

1.1.1.1 Brimonidine

- 5–Bromo–N–(4,5–dihydro–1H–imidazol–2–yl)–6–quinoxalinamine
- (5–bromoquinoxalin–6–yl)–(4,5–dihydro–1H–imidazol–2–yl)amine
- (5–bromoquinoxalin–6–yl)–2–imidazolin–2–ylamine
- 5–Bromo–6–(2–imidazolin–2–ylamino)quinoxaline
- 5–Bromo–N–(imidazolidin–2–yl)quinoxalin–6–amine
- 6–Quinoxalinamine, 5–bromo–N–(4,5–dihydro–1H–imidazol–2–yl)–(9CI) [1]

1.1.1.2 Brimonidine tartrate

- Bromo–6–(2–imidazolin–2–ylamino)quinoxaline D–tartrate
- bromo–6–(imidazolidinylideneamino)quinoxaline
- 5–bromo–6–(imidazolin–2–ylamino)quinoxalin
- (5–bromoquinoxalin–6–yl)–2–imidazolin–2–ylamine,
  2,3–dihydroxybutanedioic acid.
- 5–Bromo–6–(2–imidazolin–2–ylamino)quinoxaline tartrate
- 5–bromo–N–(4,5–dihydro–1H–imidazol–2–yl)–6–quinoxalinamine
  tartrate [2]

1.1.2 Non-proprietary names (generic)

- Brimonidine; Brimonidinum; Brimonidin; Brimonidine; Brimonidina;
  AGN 190342; BRN 0751629; UNII-E6GNX3HHTE [3].
- AGN-190342-LF; Brimonidin Tartrate; Brimonidine, Tartrate de;
  Brimonidini Tartras; Tartrato de brimonidina; UK-14304-18. 5–Bromo–
  6–(2–imidazolin–2–ylamino)quinoxaline D–tartrate [4].

1.1.3 Proprietary names (brand names)

- Alphagan, Alphagan-P, Mirvaso Lumify [5]

1.2 Formulae [5,6]

1.2.1 Empirical formula

In Table 1, the empirical formula, the molecular weight, and the CAS number of brimonidine and brimonidine D–tratrate were given.

Table 1 Empirical formula, molecular weight and CAS number.

Compounds	Empirical formula	Molecular weight	CAS number
Brimonidine	$C_{11}H_{10}BrN_5$	292.35 g/mol	59308-98-4
Brimonidine D-Tartrate	$C_{11}H_{10}BrN_5{\cdot}C_4H_6O_6$	442.23 g/mol	70359-46-5

1.2.2 Structural formula

Brimonidine tartrate and brimonidine structures [7] were drawn using MarvinSketch21.19.0 [8].

Brimonidine   Brimonidine tartrate

1.3 Elemental analysis

The theoretical elemental composition of brimonidine is shown in Table 2 [6].

Table 2 Elemental composition of Brimonidine and Brimonidine Tartrate.

Compound	Carbon	Hydrogen	Bromine	Nitrogen	Oxygen
Brimonidine	45.23%	3.45%	27.35%	23.97%	–
Brimonidine Tartrate	40.74	3.65	18.07	15.84	21.71

1.4 Appearance

Crystalline powder that is white to slightly yellowish in color [9].

2. Synthesis [8]

The procedure of preparing Brimonidine **1** or salt thereof by Bandgar et al. [10] requires the addition of Pd/C in water to a solution of 2,4-dinitroaniline **2** in methanol, and maintaining the reaction mixture under hydrogen pressure at 30–35 °C for 1 h. The Pd/C catalyst was collected by filtering at suction once the reaction was complete and the mixture was allowed to cool at room temperature. Compound **3** as 1, 2, 3-triaminobenzene dihydrochloride was produced after the organic layer was treated with hydrochloric acid. Then sodium bicarbonate solution in water and glyoxal sodium bisulfate were added to Compound **3**, and the reaction mixture was maintained at a temperature of 40–45 °C for 1 h. After the reaction is complete, the product is extracted with ethyl acetate and concentrated to yield 6-aminoquanoxaline **4**. Then, at room temperature, N-bromo-succinimide was added to the solution of compound **4** in methylene dichloride, and the reaction mixture was shaken for 2 h. The reaction mixture was then filtered and washed with water. Concentration of the organic layer resulted in the formation of 5-bromo-6- amino-quanoxaline **5**. In addition, benzoyl chloride was added to a solution of ammonium thiocyanate in acetone at room temperature with constant stirring. To complete the reaction, a 5-bromo-6-aminoquanoxaline **5** solution was added to the reaction vessel and then the reaction was refluxed for 30 min. The solution was filtered, and the solids were hydrolyzed with a solution of dilute sodium hydroxide and mixed for a further 30 min at room temperature. Dilute hydrochloric acid was then used to neutralize the mixture. To collect 5-bromo-6-thioureidoquinoxaline **6**, the reaction mixture was filtered at suction. Following that, sodium bicarbonate and dimethyl sulfate were added to the compound 6 solution in methanol with constant stirring at room temperature. For 1 h, the resultant solution was refluxed. Following that, the mixture was concentrated and water was added while stirring continuously at room temperature. Finally, the precipitated product was filtered under suction to obtain *S*-methyl-5-bromo-6-thioureidoquinoxaline **7**. Ethylenediamine-p-toluene sulfonic acid was added to a solution of *s*-methyl 5-bromo-6-thioureidoquinoxaline **7** in isopropyl alcohol with continuous stirring at room temperature. The reaction mixture that resulted was heated to reflux for 1 h. After the reaction was complete, the reaction mixture was filtered and rinsed with water to obtain Brimonidine **1** (Scheme 1).

Danilewicz et al. [11] synthesized Brimonidine using 4-nitro-1, 2-phenylenediamine as a starting material and nitro reduction to obtain 1,2,4-triaminobenzene, which was then cyclized with glyoxal and finally used bromine. Bromination results in the formation of the critical intermediate 5-bromo-6-aminoquinoxaline (**6**), which is subsequently reacted with thiophosgene to form 5-bromo-6-isothiocyanatoquinoxaline (**7**), which is finally cyclically closed with ethylenediamine to provide brimonidine 1. Brimonidine L-tartrate is formed when brimonidine **1** is treated with L-tartrate (Scheme 2).

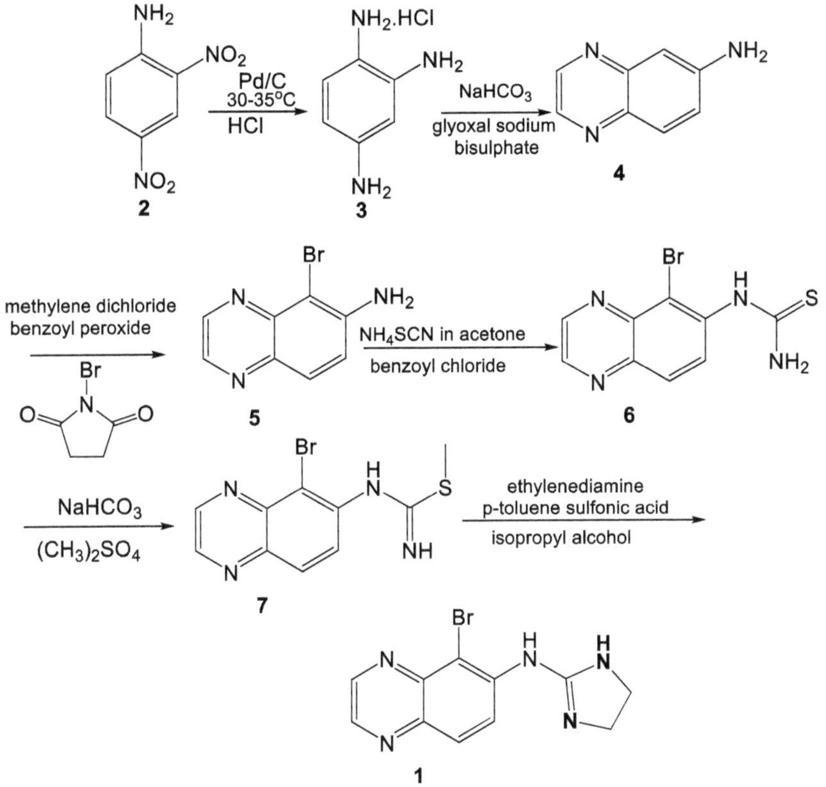

Scheme 1 Synthesis of Brimonidine using 2,4-dinitroaniline as starting martial [10].

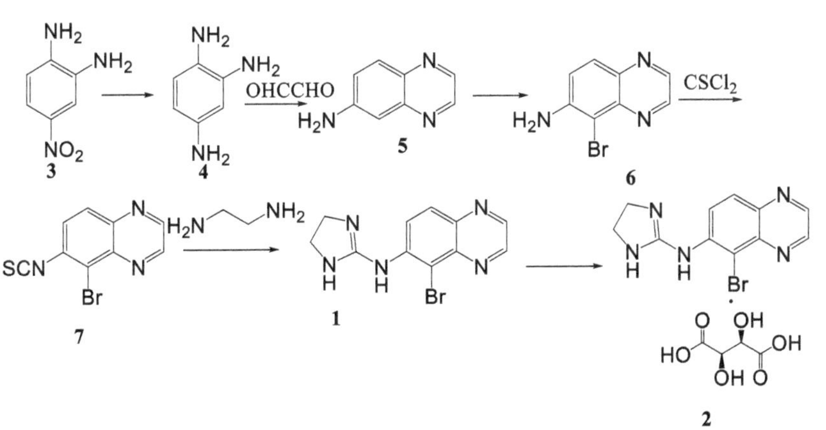

Scheme 2 Synthesis of Brimonidine using 4-nitro-1,2-phenylenediamine as starting martial [11].

Thomas et al. [12] used the same strategy methodology as Danilewicz [13] to prepare the key intermediate 6-aminoquinoxaline **5**, which reacts with ammonium thiocyanate to give 1-(5-bromoquinoxalin-6-yl)thiourea **6**, then it undergoes methylation with methyl iodide to provide 1-(5-bromoquinoxalin-6-yl)-2-methylisothiouronium **7**, that compound **7** was treated with an ethylenediamine to cyclically ring to produce Brimonidine **1**, and then Brimonidine **1** L-tartrate are produced after further reaction with tartaric acid. Among them, 5-Bromoquinoxalin-6-amine **6**, as an important intermediate of brimonidine tartrate, directly affects the tartrate industrialization process of brimonidine acid (Scheme 3).

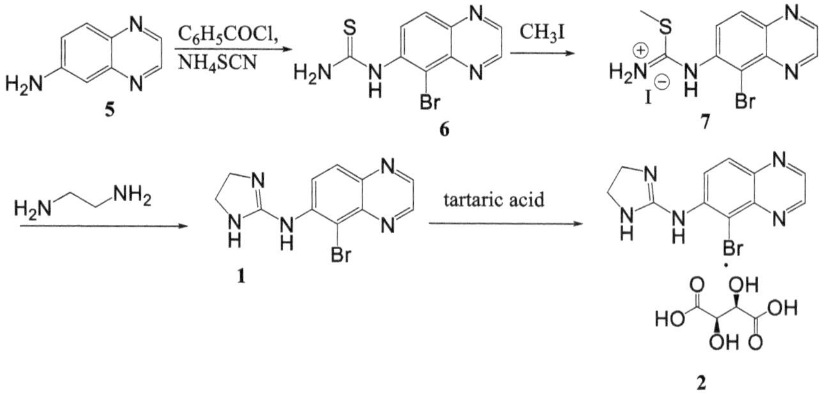

Scheme 3 Synthesis of Brimonidine using key intermediate 5-bromo-6-aminoquinoxaline [12].

In addition, Chen et al. [14], Zhang et al. [15] and Zhang et al. [16] used 1,3-dibromo-5,5-dimethylhydantoin (DBDMH) or N-bromosuccinimide amine (NBS) replaces liquid bromine as a brominating agent for bromination of 6-aminoquinoxaline. Although this method avoids the use of highly toxic liquid bromine, no bromate and polybrominated impurities are generated, but when N-bromosuccinimide (NBS) is used for bromination, the selectivity is poor and polybromination will occur. In addition, 1,3-Dibromo-5,5-dimethylhydantoin (DBDMH), used as a brominated reagent, is expensive and requires more substrates and solvents, which increases the difficulty of operation. Also, this preparation method does not meet the requirements of green chemistry (Scheme 4).

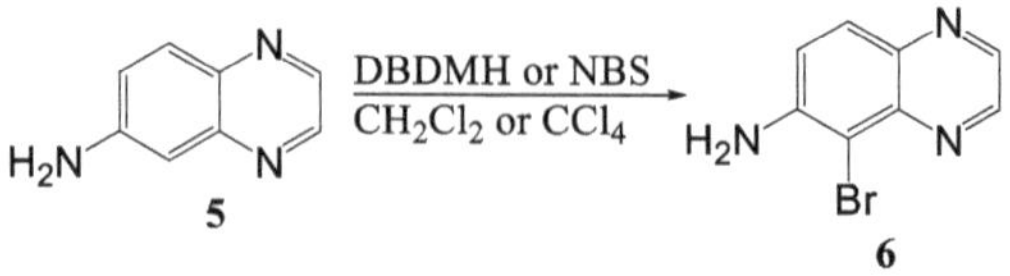

Scheme 4 Synthesis of Brimonidine using 1,3-dibromo-5,5-dimethylhydantoin (DBDMH) or N-bromosuccinimide Amine (NBS) for bromination of 6-aminoquinoxaline.

Glushkov et al. [17] developed a method for preparing brimonidine (**1**) from the reaction of amine **2** with thiophosgene **3** to produce 1-(2-aminoethyl)-3-(5-bromoquinoxalin-6-yl)thiourea **4**. Finally, compound **4** was cyclically ringed to yield Brimonidine **1** (Scheme 5).

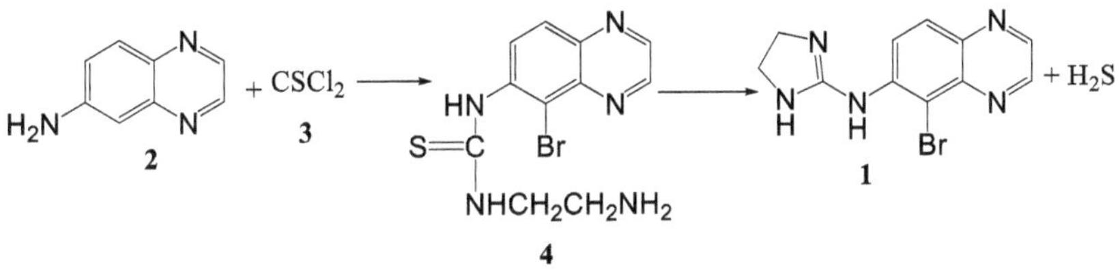

Scheme 5 Synthesis of Brimonidine using Glushkov method [18].

According to the proposed method, N-(dichloromethylene)-*N*-methyl-methanaminium **3** reacted with 5-bromo-6-isothiocyanatoquinoxaline **2** to give (E)-N'-(5-bromoquinoxalin-6-yl)-*N*,*N*-dimethylcarbamimidic chloride **8** [18], which was condensed with amine (II) in an organic solvent, for example, in chloride methylene to yield (E)-3-(2-aminoethyl)-2-(5-bromoquinoxalin-6-yl)-1,1-dimethylguanidine **9**. After extraction from methylene chloride with water, compound **9** is cyclized at $20\,°C$ to give brimonidine base, which precipitates out of solution as a crystalline precipitate, and the resulting dimethylamine dissolves in water (Scheme 6).

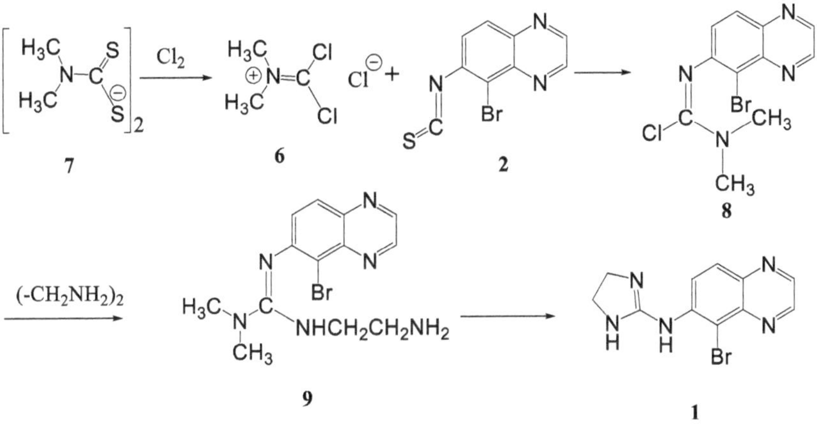

Scheme 6 Synthesis of Brimonidine using Viehe method [18].

Naik et al. [19] prepared brimonidine and brimonidine tartrate as following: at a temperature of $55–60\,°C$, N-acetyl ethylene urea **4** and 6-amino-5-bromo quanoxaline **3** were reacted to generate *N*-acetyl brimonidine **5**.

Then, acetyl brimonidine was hydrolyzed in methanolic sodium hydroxide to obtain brimonidine base **1**, which when reacted with tartaric acid in methanol yields and purifies brimonidine tartrate **2** (Scheme 7).

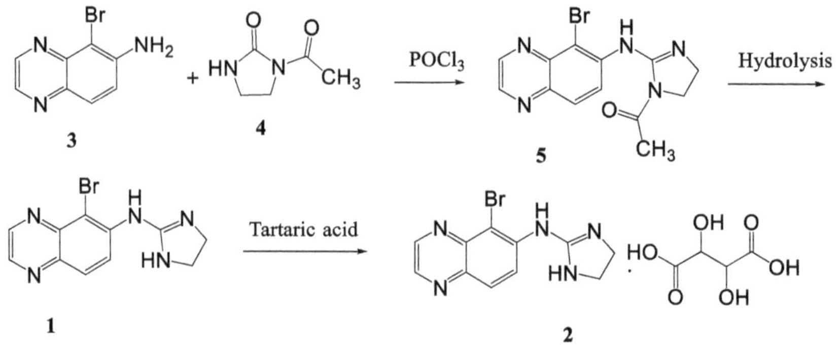

Scheme 7 Synthesis of Brimonidine using Naik method [19].

3. Physical characteristics

3.1 Dissociation constant [20]

The dissociation constant of brimonidine is shown in Table 3.

Table 3 Dissociation constant of Brimonidine and Brimonidine tartrate.

Dissociation level	pKa1	pKa2
Dissociation constant	7.64	1.82
Type (acidic/basic)	Basic	Basic

3.2 Solubility characteristics

Brimonidine tartrate dissolves in water, anhydrous ethanol and toluene are practically insoluble [21].

3.3 Specific optical rotation

The specific optical rotation of the dried brimonidine tartrate substance is in the range of +9.0 to +10.5 [21].

3.4 Density

The density of brimonidine is $1.8 \pm 0.1\,g/cm^3$ [22].

3.5 Partition coefficient

The partition coefficient of brimonidine is about 1.7 [1].

3.6 Thermal methods of analysis

3.6.1 Melting behavior

Brimonidine's melting point was determined to be 208 °C using a Büchi automatic apparatus, model B-545, which agreed with the reference melting point of 207 °C [23].

3.6.2 Differential scanning calorimetry (DSC)

Differential scanning calorimetry (DSC) data for Brimonidine was collected with a Perkin-Elmer DSC-7 differential scanning calorimeter connected to a Perkin-Elmer TAC 71DX thermal analysis controller (Perkin-Elmer, Norwalk, CT). To obtain thermograms, samples were carefully weighed (3.5 mg) into aluminium pans and heated at a rate of 10 °C/min throughout a temperature range of 40–300 °C. Pyris 2.04 (Perkin-Elmer) software was utilized for analysis.

Brimonidine's differential scanning calorimetry (DSC) curve (Fig. 1) demonstrated an endothermic reaction at 214.21 °C.

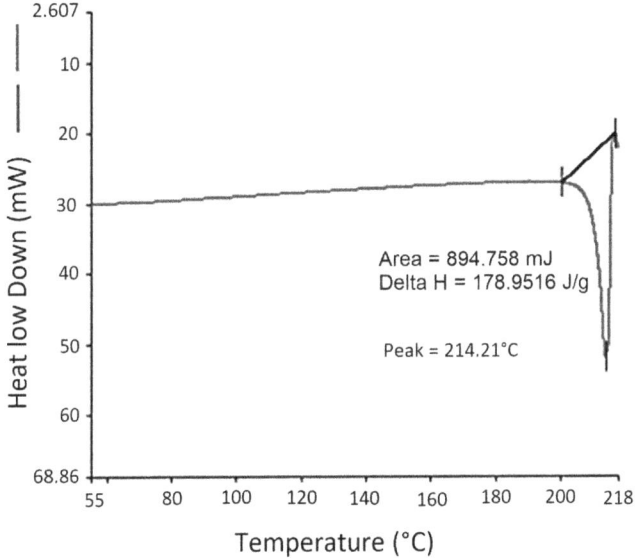

Fig. 1 Differential scanning calorimetry (DSC) curve of brimonidine.

3.6.3 Boiling point, and flash point [22]

In general, the boiling and melting temperatures of molecules with a bigger size are higher. Take into consideration the boiling temperatures of hydrocarbons that are progressively bigger. A greater number of carbons and hydrogens results in a larger surface area that may be affected by London forces, which leads to higher boiling points. Brimonidine is a solid at temperatures below zero degrees Celsius (and at atmospheric pressure) because these forces lead the brimonidine molecules to stick to one another. However, once the temperature rises beyond zero degrees, the individual molecules acquire sufficient thermal energy to break apart from one another and enter the gas phase. Brimonidine, on the other hand, does not transition out of the liquid state until it reaches a temperature of 432.6 ± 55.0 °C at 760 mmHg. This is because to the higher effect of attractive London forces, which is only feasible since the individual molecules have a larger surface area.

The flash point is an indication of the combustibility or flammability of a substance. Below the flash point, there is insufficient vapour to sustain burning. At a certain temperature over the flash point, the liquid will create sufficient vapour to sustain burning. The flash point of Brimonidine molecules is so high that it is difficult to ignite anything at temperatures lower than 215.4 ± 31.5 °C.

3.7 Spectroscopy

3.7.1 UV/vis spectroscopy

The ultraviolet absorption spectrum of brimonidine at a concentration of 20 ng/mL in methanol was recorded using a Shimadzu UV–spectrophotometer model UV-1800 equipped with 1 cm matched quartz cells. The absorption spectrum of Brimonidine solution was measured between 200 and 400 nm. Brimonidine had two maxima at 247.40 and 302.60 nm, as illustrated in Fig. 2.

3.7.2 Fluorescence spectroscopy

Fluorescence measurements were carried out on a Jasco FP-8200 Fluorescence Spectrometer (Jasco Corporation, Japan) equipped with a 150 W xenon lamp and 1 cm quartz cells. The slit widths for both the excitation and emission monochromators were set at 5.0 nm. The calibration and linearity of the instrument were frequently checked with standard quinine sulfate (0.01 μg/mL). A wavelength calibration was performed by measuring λ_{ex} at 275 nm and λ_{em} at 430 nm; no variation in the wavelength was observed. All recorded spectra were converted to ASCII format

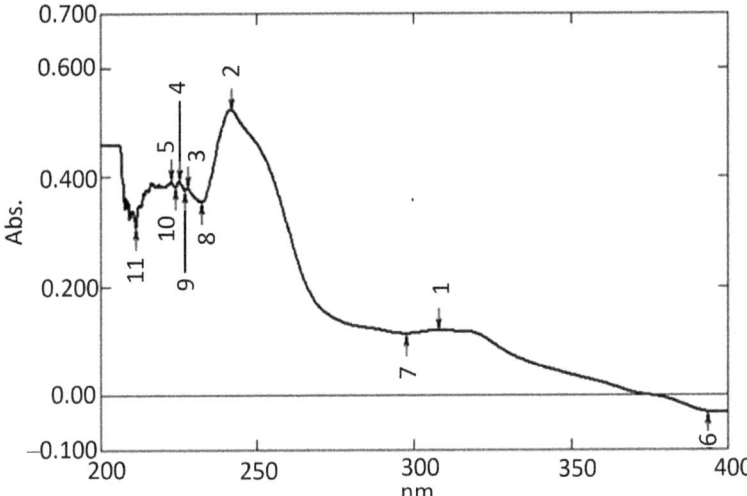

Fig. 2 UV absorption spectrum of Brimonidine in methanol.

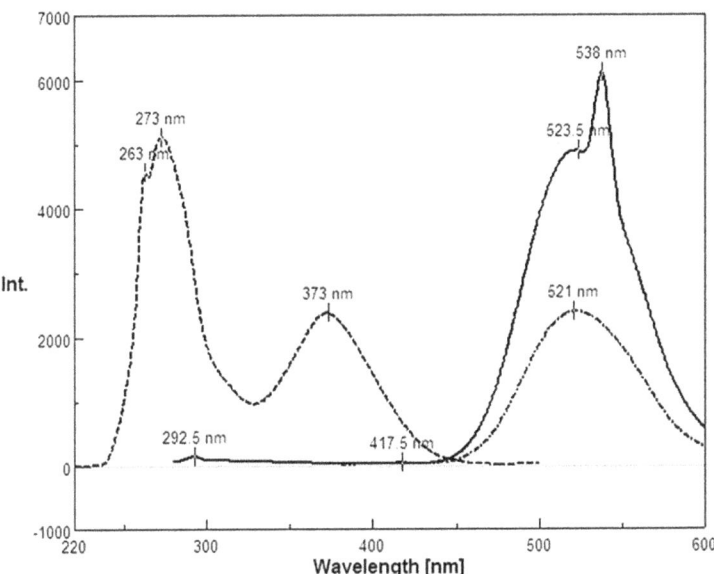

Fig. 3 Plot of Brimonidine (20 µg/mL) fluorescence emission intensity *I* vs wavelength λ.

by SpectraManager® software. A Hanna pH-Meter (Romania) was used for pH adjustments. The fluorescence spectrum is shown in Fig. 3 and Brimonidine exhibited maximum excitation at 273 and maximum emission at 538 nm.

3.7.3 Infrared spectroscopy (FTIR)

The Perkin Elmer FT-IR Spectrum BX was used to record the infrared absorption spectrum of Brimonidine as a pure powder. Fig. 4 shows the *FT-IR* spectrum of brimonidine.

The spectrum of brimonidine had the typical peaks at $1265.06 \, \text{cm}^{-1}$ (C=N stretch), $1599.22 \, \text{cm}^{-1}$ (—NH bend), $1618.18 \, \text{cm}^{-1}$ (—C=C stretch), and $1734.17 \, \text{cm}^{-1}$ (—C=O stretch).

3.7.4 Nuclear magnetic resonance spectrometry
3.7.4.1 ^{1}H NMR spectrum

^{1}H NMR spectrum of Brimonidine was scanned in DMSO-d_6 on a Brucker NMR spectrometer operating at 500 MHz (Table 4). Chemical shifts are expressed in δ-values (ppm) relative to TMS as an internal standard (Fig. 5).

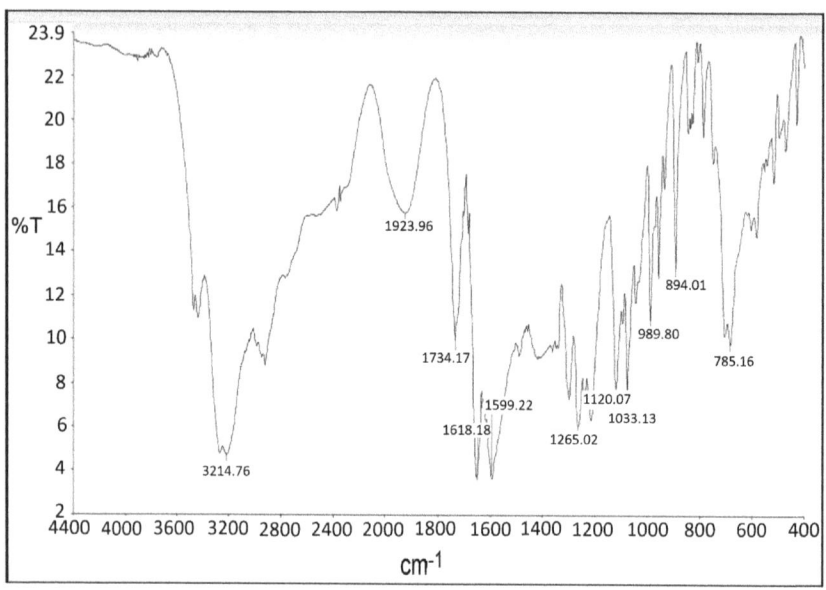

Fig. 4 Infrared absorption spectrum of Brimonidine.

Table 4 ^{1}H NMR of Brimonidine (DMSO).

Signal	Chemical shift	Splitting	Integration	Assignment
24, 25, 26 and **27**	3.49–3.62	m	4	$2 \times \underline{CH_2}$
23 and **22**	4.16	S	2	$2 \times N\underline{H}$
18	7.70–7.78	M	1	Ar$\underline{H}$
20	7.98–8.05	M	1	Ar$\underline{H}$
19	8.84–8.91	M	1	Ar$\underline{H}$
21	8.95–9.01	M	1	Ar$\underline{H}$

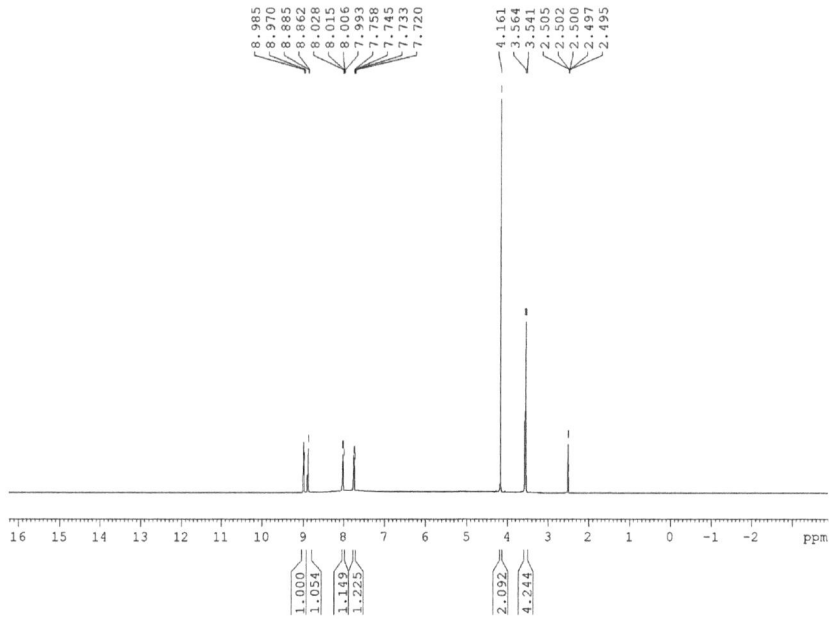

Fig. 5 ^{1}H NMR spectrum of Brimonidine.

3.7.4.2 ^{13}C NMR spectrum

^{13}C NMR spectrum of Brimonidine was scanned in DMSO-d_6 on a Brucker NMR spectrometer operating at 125 MHz (Table 5). Chemical shifts are expressed in δ–values (ppm) relative to TMS as an internal standard (Fig. 6).

Table 5 ^{13}C NMR of Brimonidine (DMSO).

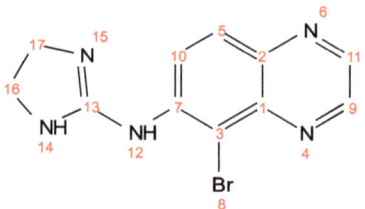

Signal	Chemical shift	Assignment
17	42.31	C̲H$_2$ (imidazolidine)
16	72.07	C̲H$_2$ (imidazolidine)
3	118.01	ArC̲ (quinoxaline)
10	128.75	C̲H (quinoxaline)
5	129.33	C̲H (quinoxaline)
1	140.52	ArC̲ (quinoxaline)
2	141.20	ArC̲ (quinoxaline)
7	144.13	N=C̲H (quinoxaline)
9	145.94	N=C̲H (quinoxaline)
11	158.40	ArN̲ (quinoxaline)
13	174.03	C̲=N (imidazolidine)

3.8 Mass spectrometry

The mass spectrum of Brimonidine ($C_{11}H_{10}BrN_5$, 292.35) was obtained using an Agilent 6320 Ion Trap Mass Spectrometer (Agilent Technologies, USA) equipped with an electrospray ionization interface (ESI). A connector was used instead of a column. The mobile phase is composed of a mixture of solvents A and B (50:50), where A is HPLC grade water, and B is acetonitrile. To prepare the compound, the solid substances were weighed out to 1 mg/mL in DMSO and then diluted with the mobile phase. The test solution was prepared by diluting the stock solutions to 10–30 μL with mobile phase. The flow rate was 0.4 mL/min and the run time was 5 min. MS parameters were optimized for each compound. The scan was in ultra-scan mode. MS2 scans were performed in the mass range of m/z 50–1000. The ESI was operated in positive mode. The source temperature was set to 350 °C nebulizer gas pressure of 55.00 psi and a dry gas flow rate of 12.00 L/min. The Brimonidine ion peak was seen at $m/z = 292$ $[M+1]^+$ as shown in Fig. 7. The fragments are also

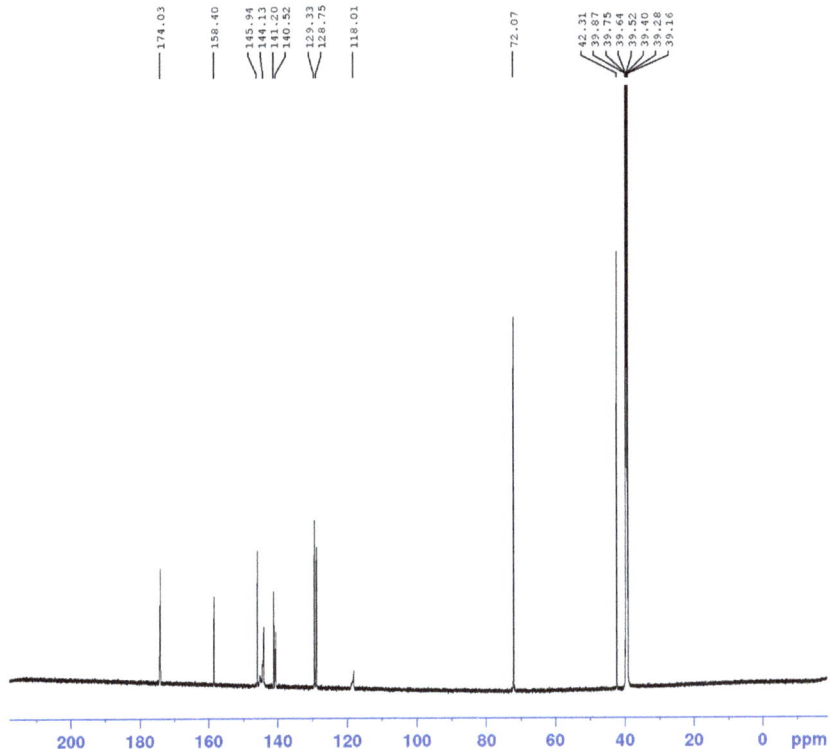

Fig. 6 ^{13}C NMR spectrum of Brimonidine.

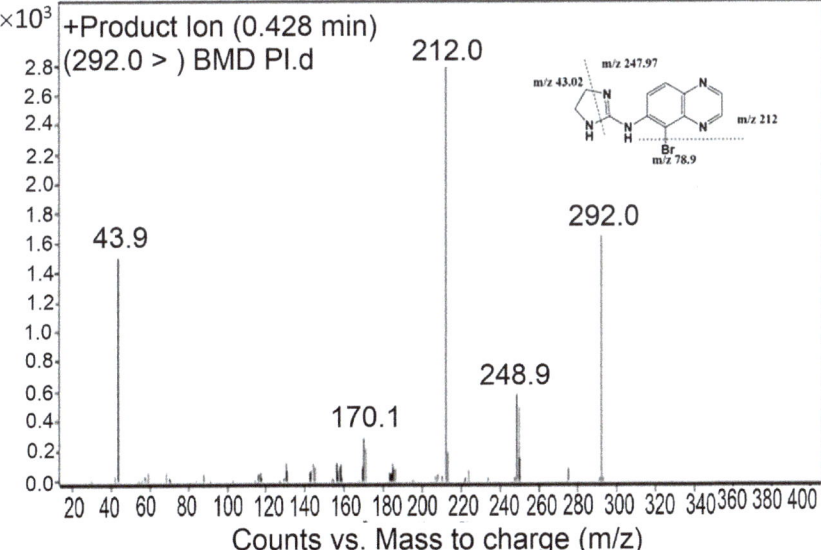

Fig. 7 Product Ion mass spectra of Brimonidine in its pure form.

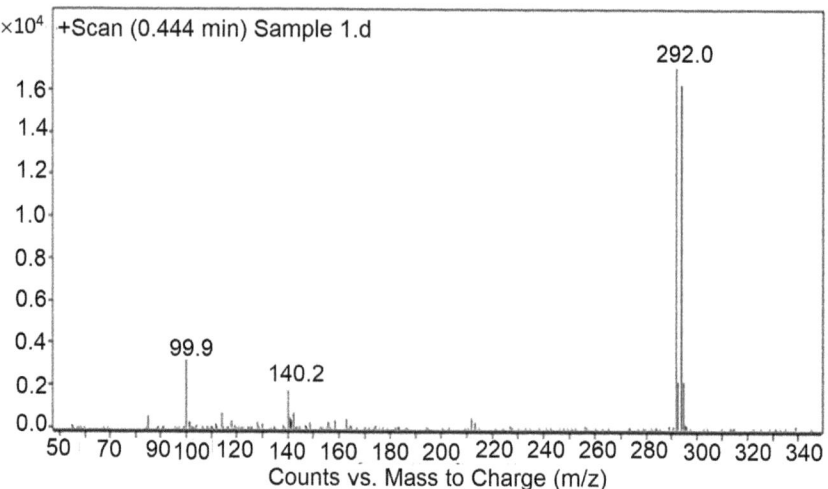

Fig. 8 Full scan mass spectra of Brimonidine in its pure form.

shown in Fig. 8. The data gathered from the MS agrees with the data mentioned in the literature [24]. Furthermore, the ESI negative ion mode is claimed to be utilized, with the key fragments (m/z) being 248.9, 212.0, 170.1, and 43.9.

4. Methods of analysis [15]

4.1 Compendial methods of analysis

4.1.1 Identification test

USP 37 [7] describes two ways to identify Brimonidine Tartrate in large quantities. These are IR absorption and a liquid chromatography system (by comparing the time the substance stays in solution).

British Pharmacopeia 2016 [21]: three methods (A, B, and C) are provided for determining the identity of Brimonidine Tartrate, namely (A) specific optical rotation, (B) infrared absorption spectroscopy, and (C) specific optical rotation. For identification, methods A and B or A and C must be used. Method B should be carried out using liquid chromatography. The sample used in this procedure has impurity E, and its area shouldn't be more than 0.10% of the area of the main peak in the standard solution.

Two limit tests for Brimonidine Tartrate-related substances were reported in USP 37 [7] utilizing LC techniques. The first test is a limit test for Brimonidine Tartrate (RS), whereas the second test is a combined test for Brimonidine Tartrate (RS) and Brimonidine Related Compound E (RS).

The percentage of each impurity in a portion of Brimonidine Tartrate was calculated by using the formula:

$$Reault = \left(\frac{r_u}{r_s}\right) \times \left(\frac{C_s}{C_u}\right) \times 100$$

where r_u denotes the peak response of each impurity in the sample solution and r_s denotes the peak response of brimonidine in the standard solution. The concentration of USP Brimonidine Tartrate RS in the standard solution is given as C_s, and the concentration of brimonidine Tartrate in the Sample solution is given as C_u.

4.1.2 Assay

The method described in USP 37 [7] for the analysis of Brimonidine Tartrate in water was based on liquid chromatography. The mobile phases are prepared in a 1-L volumetric flask based on dissolving 2.6 g of sodium 1-heptanesulfonate in 310 mL of methanol, adding 2.5 mL of triethylamine and 7.5 mL of glacial acetic acid, and diluting with water to volume. The LC method was used (4.6-mm, 25-cm; 5-m packing L1), and the stationary phases were prepared in Furthermore, the flow rate is set to 1 mL/min, the injection volume is 20 L, the column temperature is 30 °C, and the wavelength of the detector is set to UV 264 nm. The Standard solution contains 1.3 mg/mL USP Brimonidine Tartrate RS in water, whereas the sample solution contains 1.3 mg/mL of Brimonidine Tartrate in water.

The British Pharmacopeia 2016 [21] the potentiometry method was employed to determine the concentration of Brimonidine in bulk. The sample (0.350 g) was sonicated until completely dissolved in 70 mL of anhydrous acetic acid R and then titrated with 0.1 M perchloric acid. And the end point was potentiometrically calculated. 1 mL of 0.1 M perchloric acid is equivalent to 44.22 mg of $C_{15}H_{16}BrN_5O_6$.

4.1.3 Impurities

Detection of other detectable impurities would be identified by one or more of the tests described in the monograph if they were present in sufficient quantities (Table 6). They are constrained by the general acceptance criterion for other/unspecified impurities and/or the general monograph on substances for pharmaceutical use, both of which are applicable in certain circumstances (2034). As a result, it is not important to detect these impurities in order to demonstrate conformity with the regulations 5.10, Pharmaceutical Substance Impurity Control: A, B, C, D, E, F, and G [21].

Table 6 Impurity control in pharmaceutical substances [7].

A. N-(imidazolidin-2-ylidene)quinoxalin-6-amine

B. 5-bromoquinoxalin-6-amine

C. quinoxalin-6-amine

D. 1-(5-bromoquinoxalin-6-yl)thiourea

E. 2-(5-bromoquinoxalin-6-yl)guanidine

F. 5-bromo-N-(1H-imidazol-2-yl)quinoxalin-6-amine

G. 1-(2-aminoethyl)-3-(5-bromoquinoxalin-6-yl)urea

4.2 Electrochemical methods of analysis

Radulovic et al. [25] conducted a comprehensive electrochemical method for the determination of brimonidine at a boron doped diamond electrode (BDDE) using cyclic voltammetry (CV) and square-wave voltammetry (SWV) of various amounts in sulfuric acid (pH ranged from 0.6 to 1.6), as well as in BR buffer (pH ranged from 2.0 to 9.0). The reduction of brimonidine was shown to proceed via a one-step quasi-reversible mechanism in which two electrons and two protons are transferred. The quinoxaline ring reduction process was discovered to be identical to that of the other quinoxaline derivatives. It was discovered that the electrode process was diffusion-controlled in acidic media, whereas a considerable degree of adsorption was found in alkaline media. Two advanced voltammetric methods, namely; differential pulse voltammetry (DPV) and square wave voltammetry (SWV), were fully validated. Linearity was achieved for DPV at concentrations ranging from 2×10^{-6} to 3×10^{-5} M (LOD $= 6.31 \times 10^{-7}$ M, LOQ $= 2.1 \times 10^{-6}$ M) and for SWV at concentrations ranging from 5×10^{-7} to 1.5×10^{-5} M (LOD $= 1.28 \times 10^{-7}$ M, LOQ $= 4.28 \times 10^{-7}$ M). The procedures were used to determine the concentration of brimonidine in a pharmaceutical dose form and eye drops.

Aleksić et al. [26] used a glassy carbon electrode (GCE) to explore the electrochemical behavior of brimonidine. They used cyclic voltammetry, differential pulse voltammetry, and square wave voltammetry (GCE). In acid and neutral media, the reduction of BRIM proceeds in a one-step quasi-reversible reaction, achieving complete reversibility in alkaline solutions. The reduction reaction involves the transfer of two electrons and two protons to the quinoxaline moiety's pyrazine ring, resulting in the formation of a dihydro-derivative. Additionally, brimonidine is permanently oxidized via the transfer of an electron and a proton to the secondary amine moiety. They examined the impacts of the electrolyte solution's pH, scan rate, and BRIM concentration. The electrode process was found to be regulated by adsorption at pH > 6, and the total surface concentration of brimonidine adsorbed onto the GCE surface at pH 7 was determined to be BRIM $= 1.35 \times 10^{-10}$ mol cm^{-2}. Based on the findings of this investigation, a differential pulse voltammetric approach was devised, verified, and recommended for quick electroanalytic determination of brimonidine at low concentrations. Within the concentration range of 5×10^{-7} to 5×10^{-6} M, linearity was attained with a LOD of 1.6×10^{-7} M and a LOQ of 5.3×10^{-7} M. The technique was used to determine the concentration of brimonidine in a medicinal dose form, eye drops.

Radulović et al. [27] established a voltammetric technique for determining brimonidine in deproteinized aqueous humor, reducing sample preparation for stability experiments. On the basis of characteristic oxidation peaks, the differential pulse voltammetric (DPV) approach using boron doped diamond electrode (BDDE) was devised and successfully utilized. The linearity range of brimonidine was between 5×10^{-6} and 5.0×10^{-5} M, while the limit of detection and quantitation were 1.94×10^{-6} and 6.46×10^{-6} M, respectively. Precision and accuracy were assessed intra- and inter-day, and all results were consistent with validation ICH recommendations. The best results for short-term stability were obtained at a concentration level of 3.0×10^{-5} M, with a variation of +1.86% between the starting and post-storage concentrations. A study of long-term stability at two concentrations of 3.0×10^{-5} M and 5.0×10^{-5} M revealed variations of +1.63% and +3.56%, respectively.

An examination of the freeze-thaw stability of samples found that they could be frozen just once. The improved sensitivity of the DPV/BDDE method obtained through modification was sufficient for the determination of immeasurable brimonidine concentrations in native, untreated aqueous humor at concentrations of 6 or 12 nmol/0.1 mL with acceptable accuracy (up to +7.5%). The sensitivity was limited by the nature of the process—the irreversible one-electron oxidation voltammetric peak of brimonidine. Only electrochemical pretreatment of the BDD electrode before each measurement greatly accelerated the entire procedure. The suggested method has the benefits of simplicity, rapid performance, high specificity and selectivity, and acceptable accuracy, and it does not require chemical modification of BDDE.

Abou Al Alamein et al. [28] developed a carbon paste electrode that had iron (III) oxide nanoparticles (MCPE) for the simultaneous quantitation of timolol maleate and brimonidine tartrate utilizing chemometrics. A voltammetric method was developed for the simultaneous detection of timolol maleate and brimonidine tartrate in the commercialized ophthalmic formulation. Iron oxide nanoparticles are incorporated into the carbon paste electrode to make the electrode's properties better and its sensitivity higher. Cyclic voltammetry (CV), DPV, and SWV techniques were used to determine the electrochemical oxidation of the two medications. The differential pulse voltammetry and square wave voltammetry procedures were developed for the quantitative voltammetric determination of timolol maleate and brimonidine tartrate in their pure forms and pharmaceutical formulations. Square wave voltammetry had a minimum detectability (LOD) of 1.31×10^{-6} µg/mL for brimonidine tartrate and 1.37×10^{-5} µg/mL for timolol maleate, and a limit of quantitation (LOQ) of 3.97×10^{-6} µg/mL for brimonidine tartrate and 4.16×10^{-5} µg/mL for timolol maleate. The resolution of the overlapping peaks of both substances and their detection in mixtures were determined using two different chemometric methods: principal component regression (PCR) and partial least squares (PLS). The PLS method produced the best results. The suggested method was validated and compared to a previously published HPLC method.

4.3 Spectroscopic methods of analysis

4.3.1 Spectrophotometry

Annapurna [29] developed multicomponent spectrophotometric techniques for determining timolol maleate and brimonidine tartrate in pharmaceutical formulations simultaneously (eye drops). The three approaches were developed using a phosphate buffer with a pH of 7.0. The absorption maxima of timolol maleate and brimonidine tartrate were determined to be 295 and 247 nm, respectively. For timolol maleate, linearity was observed between 1 and 120 µg/mL and for brimonidine tartrate, linearity was observed between 1 and 60 µg/mL, and the three procedures were validated according to ICH requirements. There are three methods available for determining timolol maleate and brimonidine tartrate concentrations in eye drops.

Damle et al. [30] established a process for estimating the Brimonidine Tartrate and Timolol Maleate combined concentration. A derivative spectroscopy method was used to reduce spectral interference in the determination of Brimonidine Tartrate and Timolol Maleate by measuring absorbance at 251.5 and 228 nm, respectively. In the first derivative spectroscopy mode, validation was performed according to ICH criteria, and linearity data revealed the LOD and LOQ to be 0.33 and 1.00 µg/mL, respectively, for Brimonidine Tartrate, and 0.13 and 0.40 µg/mL, respectively, for Timolol Maleate.

Annapurna et al. [31] presented spectrophotometric verified techniques for Brimonidine and Timolol simultaneously in ocular formulations. Brimonidine and timolol in borate buffer were determined simultaneously using the equation method and Q-Analysis. Both Timolol and Brimonidine exhibited linearity at concentrations of 1–60 and 1–40 µg/mL, respectively. The two procedures were tested and proven to work, and they could be used to detect the Brimonidine and Timolol in eye drops simultaneously.

Four spectrophotometric techniques for the assay of Brimonidine tartrate in pharmaceutical dosage forms were established by Annapurna et al. [32]. The first approach (Method A) employed phosphate buffer at pH 6.0 to produce absorption maxima at 247 nm, whereas the second method (Method B) utilized borate buffer at pH 9.0 (λ_{max} 257 nm). The concentration range of 0.1–50 µg/mL was investigated for both methods with regression equations of $0.0656 \times - 0.0121$ and $0.0606 \times + 0.0038$ for method A and B, respectively. Initially, first derivative spectrophotometric methods (C and D) were developed in phosphate buffer pH 5.0 and sodium acetate pH 4.0, respectively, wherein Brimonidine tartrate conforms Beer Lambert's law at concentrations of 1–40 and 0.2–30 mg/mL, respectively, with regression equations of 0.0034×-0.0007 and $0.0064 \times +0.0004$, respectively. These spectrophotometric methods were good for determining the quantity of Brimonidine tartrate in drug formulations.

de Souza et al. [33] developed and validated a spectrophotometric method for determining the concentration of brimonidine incorporated into and discharged from chitosan implants. Brimonidine was identified at a wavelength of 258 nm. The analytical method was linear throughout a concentration range of 0.5–15 µg/mL brimonidine, with an average recovery of 96–102%. The theoretical limit of quantification was 0.015 µg/mL. The validated approach was used to quantify the brimonidine contained in chitosan implants, demonstrating that the drug was distributed uniformly throughout these implantable devices. Finally, the spectrophotometric approach demonstrated that the brimonidine integrated into the chitosan was released in a regulated and extended manner in vitro.

Popaniya and Patel [34] developed two UV spectroscopic approaches for the simultaneous determination of brimonidine tartrate and timolol maleate, namely simultaneous equation (method 1) and Q-absorbance ratio (method 2). Both compounds exhibited excellent linearity in method 1 over the concentration ranges of 2–14 and 5–35 µg/mL of Brimonidine tartrate and timolol maleate at 247.0 nm ($r^2 = 0.999$) and 295.0 nm ($r^2 = 0.998$). Method 2 entails the formation of a Q-absorbance equation using the absorptivity values at 266.0 nm (isoabsorptive point) and 295.0 nm (λ_{max} of timolol maleate) ($r^2 = 0.998$).Beer's law was followed over the concentration ranges of 2–14 and 5–35 µg/mL for brimonidine tartrate and timolol maleate, respectively. In ocular formulations, the two techniques have the potential for simultaneous measurement of Brimonidine tartrate and timolol maleate, and for collapse detection in aqueous solutions.

Rizk et al. [35] established two spectrophotometry assays for the evaluation of a binary mixture including timolol maleate and brimonidine tartrate in the presence of benzalkonium chloride as an eye drop preservative. Utilizing UV spectrophotometry, the first derivative was determined with zero-crossing observations at 313 nm for timolol maleate and 386 nm for brimonidine tartrate. The second approach was based on the first derivative of the ratio spectrum, with zero-crossing readings at 313 and 391 nm for timolol maleate and brimonidine tartrate, respectively, using 0.25 µg/mL benzalkonium chloride as a divisor. Linearity was observed over the concentration ranges of 5–85 and 2–35 µg/mL for timolol maleate and brimonidine tartrate, respectively.

Jain et al. [36] described a first-order UV-derivative spectrophotometric technique for estimating brimonidine tartrate in bulk and pharmaceutical formulations. The maximum absorption wavelength of brimonidin tartrate in methanol and water was determined to be 247 nm. Derivatization of the same spectrum to a first order derivative revealed the largest amplitude of the trough at 262 nm. In the concentration range of 2–12 µg/mL, the substance exhibited linearity with a correlation coefficient of 0.998. The procedure was applied to pharmaceutical formulations, and the average recovery was found to be 99.58%, which is consistent with the label claim.

Elzanfaly et al. [37] developed a spectrophotometric method for the simultaneous determination of brimonidine and timolol with interfering spectra in binary mixtures without prior separation. It is based on a straightforward modification of the ratio subtraction technique. The mean percentage recoveries of brimonidine and timolol in bulk were 100.40 ± 2.29 and 101.23 ± 1.30, respectively, and the mean percentage recoveries in their pharmaceutical formulations were 101.08 ± 0.44 and 100.66 ± 0.52, respectively. This ratio difference method was validated to USP standards and it is suitable for routine quality control testing.

Jadhav et al. [38] developed a technique for the determination of Brinzolamide and Brimonidine in bulk and ophthalmic formulations simultaneously. In methanol, the absorbance was measured at two wavelengths: 252.40 nm for Brinzolamide and 246 nm for Brimonidine. At their respective maximums, Brinzolamide and Brimonidine achieved linearity in the concentration ranges of 5–35 and 3–18 µg/mL, respectively.

4.3.2 Spectrofluorimetry

Ibrahim et al. [39] developed and validated two techniques for measuring the pure state and pharmaceutical formulations of brimonidine tartrate (BT). The proposed method was based on coupling between the drug being determined and 4-chloro-7-nitro-2,1,3-benzoxadiazole (NBD-Cl) in borate buffer (pH 8.5) at 70 °C followed by detection of the reaction product spectrophotometrically at 407 nm (method I) or spectrofluorimetrically at 528 nm following excitation at 460 nm (method II). Calibration graphs for both methods (I and II) exhibited lower detection limits of 0.21 and 0.03, and lower quantification limits of 0.65 and 0.09 µg/mL for the concentration ranges of 1.0–16.0 and 0.1–4.0 µg/mL, respectively. The recovery of commercial ophthalmic solution was found to be $99.50 \pm 1.00\%$ and $100.13 \pm 0.71\%$ for methods I and II, respectively. A stability study of brimonidine tartrate was carried out under the various circumstances indicated in the ICH, including alkaline, acidic, oxidative, and photolytic destruction. Additionally, the kinetics of the drug's oxidative degradation were studied. The apparent first-order reaction rate constants, half-lives, and Arrhenius equation were also studied.

Sunitha et al. [40] developed a spectrofluorimetric method for determining the concentration of brimonidine tartrate in pure and eye drops. Linearity was observed at concentrations ranging from 0.2 to 3.0 µg/mL in dimethyl formamide as the solvent at an emission wavelength (λ_{em}) of 530 nm following an excitation wavelength (λ_{ex}) of 389 nm, with a correlation coefficient of 0.998. This method had a limit of detection and a limit of quantification of 22.0 and 72.0 ng/mL, respectively. For accuracy and precision studies, the percentage relative standard deviation values were found to be less than 2. The approach was used to conduct routine quality control analyses on the brimonidine tartrate contained in the eye drops.

4.4 Chromatographic methods of analysis

4.4.1 Electrophoresis

Tzovolou et al. [41] developed a capillary electrophoresis method for determining the amounts of brimonidine in the aqueous humor of the eye and blood serum, as well as their relationship to the drug's efficacy in lowering intraocular pressure (IOP). Brimonidine was analyzed using capillary zone electrophoresis with a buffer concentration of 20 mM borate, pH 9.3, and a detection wavelength of 255 nm. Brimonidine concentrations in aqueous humor and blood serum were evaluated in seven patients admitted for cataract extraction following ocular administration of the ophthalmic Alphagan™ solution. Brimonidine levels and IOP values were monitored for a 24-h period. Alphagan™ administration resulted in a considerable reduction of IOP from 30 min to 4–5 h, followed by a gradual increase until 24 h, at which point the mean IOP value recovered to pre-administration levels. The IOP drop was proportional to the level of brimonidine in the aqueous humor, which reached a maximum (80–100%) within 1–3 h. After 4–5 h, the solution achieved a 50% concentration, but it reaches its minimum concentration after 12 h. Serum levels reached a maximum in 3–4 h, a 50% decline in 12 h, and a low level in 24 h. The authors concluded that when brimonidine is administered at a concentration of 50% of the maximal concentration found in aqueous humor, i.e., during a 4–6 h period, it can dramatically reduce IOP in patients. Given that the level of brimonidine in blood serum achieved its maximal value at this point, administration of brimonidine every 6 h may be used to maintain continuous IOP reduction.

4.4.2 High performance liquid chromatography: (USP)

Summary of Chromatographic Methods of Brimonidine.

Sample	Method	Mobile phase	Stationary phase	Linearity range	Reference
Bulk drug and in ophthalmic formulation	RP–HPLC	Mixture citric acid monohydrate buffer:water:methanol (30:50:20 vol/vol/v) and pH 3 was maintained by using trimethylamine at a flow rate of 1 mL/min Detection wavelength was 246 nm	C18 column was (250 mm × 4.6 mm, 5 μm)	40–80 μg/mL The LOD and LOQ was 1.47 and 4.47 μg/mL	[42]
Bulk of six drugs	HPLC	Phosphate buffer pH 5: acetonitrile (78:22) at 1 mL/min 254 nm detecting wavelength	Thermo Hypersil BDS C18 column (4.6 × 250 mm, 5 μm)	2–70 μg/mL	[43]
Suspension	RP–HPLC	Methanol: 0.01 M ammonium acetate buffer (49.5:50.5, vol/vol) pH was adjusted to 3.8 by adding acetic acid, and flow rate was 1.1 L/min, it was a detected at 260 nm	C_{18} column (250 × 4.6 mm, 5 μm)	1–7 μg/mL	[44]
An Ophthalmic Dosage Form	RP–HPLC	Phosphate buffer (pH 6.6): acetonitrile:methanol (45:15:40) delivered at 1.0 mL/min with detection wavelength of 254 nm	Phenomenex C18 (5 μm, 250 × 4.6 mm) column	50–1600 ng/mL	[45]
Eye drops	HPLC	Acetonitrile: 25 mM phosphate buffer, pH 4.0 (50:50, vol/vol) at 1.2 mL/min with UV detection at 210 nm	A BDS Hypersil phenyl column	2–80 μg/mL	[46]

Continued

—cont'd

Sample	Method	Mobile phase	Stationary phase	Linearity range	Reference
Ophthalmic formulation	HPLC	Citric acid monohydrate buffer: water:methanol (30:50:20 vol/vol/v) with triethylamine to adjusted pH 3	Kromasil C 18 (250 mm × 4.6 mm i.d., 5 µm particle size) column	The LOD and LOQ were 1.47 and 4.47 µg/mL, respectively.	[47]
Bulk and ophthalmic dosage	HPLC	Methanol/phosphate buffer solution 65:35 at pH 4 with flow rate 1.0 mL/min detection was 260 nm	C18 (5 µm, 250 × 4.6 mm) column	20–100 µg/mL	[48]
combined ophthalmic dosage	HPLC	a single mobile phase solution of acetonitrile:0.05M phosphate (30:70), pH 3.5, and wavelength 220 num	C18 (5 µm, 250 × 4.6 mm) column	2–20 µg/mL	[49]
Bulk drug and in ophthalmic formulation	HPLC	Triethylamine was used to maintain a pH of 3 in the mobile phase citric acid monohydrate buffer:water:methanol (30:50:20 vol/vol/v). The flow rate was set at 1.0 mL per minute. Elute was found at a wavelength of 246 nm	Kromasil C 18 (250 mm × 4.6 mm i.d., 5 µm particle size) column	40–80 µg/mL	[42]
Ophthalmic formulation	HPLC	Citric acid monohydrate buffer: water: methanol (30:50:20). The buffer's pH was kept at 3 by triethylamine. 1.0 mL/min flow rate Elute was detected at 246 nm	Kromasil C-18 column (250 mm × 4.6 mm i.d., 5 µm particle size)	40–80 µg/mL	[50]

Sample	Method	Mobile phase	Stationary phase	Linearity range	Reference
Ophthalmic formulation	HPLC	Methanol: 0.01 M ammonium acetate buffer (49.5:50.5, vol/vol) pH 3.8 with acetic acid at 1.1 mL/min at 260 nm	C18 column (250 × 4.6 mm, 5 μm)	0.2–1.4 μg/mL	[51]
Ophthalmic formulation	HPLC	Phosphate buffer (pH 6.6): acetonitrile: methanol (45:15:40) pumped at a flow rate of 1.0 mL/min and detected at a wavelength of 254 nm	Phenomenex C18 (5 μm, 250 × 4.6 mm) column	50–1600 ng/mL	[52]
Commercial samples of single-ingredient ophthalmic solutions	HPLC	1.2 mL/min of acetonitrile: 25 mM phosphate buffer, pH 4.0 (50:50, v/v) with UV detection at 210 nm	BDS Hypersil Ph column	2.0–80.0 μg/mL	[46]
Pharmaceutical formulations	HPLC	1 mL/min of a mixture solution of phosphate buffer (10 mM, pH3.5) containing 0.5% triethylamine and methanol (85:15, v/v) with a diode array detector at 246 nm	A Diamonsil C18 column (150 mm × 4.6 mm, 5 μm)	0.01–50 μg/mL	[53]
Combigan Eye Drops	HPLC	1.5 mL/min of Ammonium acetate (pH 5.0, 0.01 M)—Methanol (40:60, V/V) with UV detection at 254.0 nm for Brimonidine Tartrate and 300.0 nm for Timolol Maleate	BDS HYPERSIL Cyano column (250 × 4.6mm, 5 μ)	(4–24 and 10–60 μg/mL) for Brimonidine Tartrate and Timolol, respectively	[54]

Continued

Sample	Method	Mobile phase	Stationary phase	Linearity range	Reference
Ophthalmic dosage	HPLC	Acetonitrile: 0.05 M phosphate buffer at (30:70) at pH 3.5 and wavelength of 220 nm	Promosil C18 column	1.25–25 μg/mL for timolol, 4–80 μg/mL for dorzolamide, 5–50 μg/mL for brinzolamide and 2–20 μg/mL for brimonidine	[55]
Combined dosage form	RP-HPLC	Potassium phosphate, pH 3.0 to acetonitrile (60:40 v/v) at flow rate 1 mL/min with the detection wavelength was 225 nm	Zorbax SB C18 (250 mm × 4.6 mm × 2.6 μm)	2–6 μg/mL for brimonidine and 10–30 μg/mL for brinzolamide	[56]
Eye drops	HPLC	The flow rate was 0.6 mL/min at 50 °C in gradient manner for the mobile phase consisted of (A) triethylamine (TEA), (0.5% v/v, pH 4.5) and (B) acetonitrile: Gradient Method: 0.0 min (50% A:50% B)–320 nm 3.5 min (80% A: 20% B)–320 nm 4.0 min (80% A: 20% B)–210 nm 6.0 min (80% A: 20% B)–210 nm 6.5 min (50% A: 50% B)–210 nm 8.0 min (80% A: 20% B)–210 nm	Phenomenex Kinetex C18 (50 × 4.6 mm, 2.6 μm)	(1–100 and 2.25–225 μg/L) for Brimonidine Tartrate and Timolol respectively.	[57]
Eye drops	A reversed phase LC	The flow rate was 1 mL/min of Octane 1-sulfonic acid sodium salt (0.02 M) (pH 3.5 ± 0.05): acetonitrile (64:36 v/v) With UV detection at 254 nm	An Inertsil ODS 3 V column (C18)		[58]

4.4.3 Liquid chromatography–mass spectrometry–mass spectrometry (LC/MS)

Hassib et al. [59] developed a liquid chromatography–mass spectrometry–mass spectrometry method for the simultaneous detection of an anti-glaucoma ß-blocker, timolol maleate (TIM), and other anti-glaucoma drugs of different classes, namely dorzolamide hydrochloride (DOR), brinzolamide (BRZ), and brimonidine tartrate (BRM) in rabbit aqueous using eslicarbazepine as an internal standard (IS). Purification and pre-concentration of analytes from the rabbit AH matrix were accomplished via liquid-liquid extraction. On an INERTSIL® C18 ODS-3 column (150 mm 4.6 mm, 3.5 m), a mobile phase of 10 mM ammonium formate (pH 7), methanol, and acetonitrile (5:50:45, v/v/v) was used in isocratic mode at a flow rate of 0.8 mL/min. Before the detection using multiple reaction monitoring (MRM) with an electrospray ionization source at the following transitions, the technique was operated in the positive ionization mode: m/z 317.2 → 261.0 for TIM, m/z 325.1 → 199.0 for DOR, m/z 384.2 → 281.0 for BRZ, m/z 292.1 → 212.0 for Brimonidine tartrate, and m/z 255.0 → 237.0 for IS. The separation took only 3 min, and the lowest limit of quantification (LLOQ) for all cited substances was 50 ng/mL. The bioanalytical method was thoroughly validated in accordance with US-FDA and EMA criteria. The standard calibration curves for all medications were linear in the range of 50–5000 ng/mL, with a good mean regression coefficient for all medications.

Using brimonidine-d(4) as an internal standard (IS), Jiang et al. [24] validated an LC/MS/MS technique for quantifying brimonidine in ocular tissues and fluids. Brimonidine was extracted from the retina followed by sonication and vortex, iris/ciliary body and vitreous mood samples with a solution of acetonitrile: water (1:1). After dilution with acetonitrile, the watery humor, iris/ciliary body, retina, and vitreous humor samples were analyzed by reverse-phase HPLC using isocratic conditions. A multiple-reaction monitoring (MRM) in the positive electrospray mode was used to assess brimonidine (m/z transition: 292 → 212) and the internal standard (m/z transition: 296 → 216) on a 4000 Q TRAP® instrument. Analysis was completed in less than 2.0 min for each sample. The brimonidine calibration curves (1–1000 ng/mL) were built with a linear regression with a weight of $1/\times 2$. For brimonidine, the lowest limit of quantification in aqueous humor was 1.0 ng/mL, whereas in the ciliary body, iris, retina, and vitreous humor, it was 10, 12.5, and 1.6 ng/g, respectively. The method was accurate and precise, and it can be used to measure the quantity of brimonidine is in the fluids and tissues of the eye.

Tang et al. [60] used a microdialysis approach to study the pharmacokinetics of brimonidine tartrate in situ gel in the anterior chamber of the rabbit eye, and samples were evaluated using HPLC–MS/MS. For brimonidine and the internal standard, it was monitored in ESI mode during transitions $291.9 \rightarrow 212.0$ and $296.0 \rightarrow 216.0$, respectively. With a flow rate of $0.4\,mL/min$, the mobile phase was composed of acetonitrile and 0.1% aqueous formic acid (50:50, v/v). The method demonstrated an excellent linear correlation in microdialysis solutions between 5 and $5000\,ng/mL$, and the inter- and intra-day precision (relative standard deviation) was less than 4%. A pharmacokinetic analysis showed that the AUC (0–t) of in situ gel was 3.5 times that of eyedrops. This means that in situ gel makes brimonidine much more bioavailable.

5. Stability

Radulovic et al. [61] developed a voltammetric method for determining brimonidine in deproteinized aqueous humor using a DPV method with a boron-doped diamond electrode (BDDE). The linearity range of the brimonidine samples was between 5.0×10^{-6} and $5.0 \times 10^{-5}\,M$, and the limits of detection and quantitation were 1.94×10^{-6} and $6.46 \times 10^{-6}\,M$, respectively. The precision and accuracy of intra-day and inter-day measurements were examined and validated according to the ICH recommendations. The best short-term stability testing used a concentration of $3.0 \times 10^{-5}\,M$, with a variation of +1.86% between initial and post-storage concentrations. Deviations of +1.63% and +3.56% were found during long-term stability studies for 3.0×10^{-5} and $5.0 \times 10^{-5}\,M$ concentrations, respectively. The accuracy of the DPV/BDDE methods for quantifying the BRIM quantities in native, untreated aqueous humor was improved when modifications were made to the procedure (reaching +7.5% accuracy for quantities of 6 or $12\,nmol/0.1\,mL$ aqueous humor). Due to the fact that the nature of the process is irreversible, and it has an irreversible one electron oxidation voltammetric peak for BRIM, the sensitivity was limited. Pre-treatment of the BDD electrode with electrochemical means increased the speed of the entire measurement process. The suggested approach is simple, short time performance, and has the ability to achieve a high degree of specificity and selectivity, as well as good accuracy, all without the need for chemical modification of BDDE.

6. Clinical pharmacology
6.1 Mechanism of action

Brimonidine is an alpha-2 adrenergic agonist with a high degree of selectivity. Brimonidine gel applied topically may diminish erythema via direct vasoconstriction [62]. The liver substantially metabolizes brimonidine. Urinary excretion is the primary route of brimonidine and its metabolites' elimination [62].

The drug Brimonidine Tartrate Ophthalmic Solution (0.15%) is an alpha-2 adrenergic receptor agonist. Studies in animals and people show that brimonidine tartrate acts in two ways: by lowering the amount of aqueous humor that is produced and increasing the outflow of fluid from the eye [63].

6.2 Pharmacodynamics

Brimonidine Tartrate Ophthalmic Solution (0.15%) produces a maximum ocular hypotensive response 2 h after administration. Increased IOP is a significant risk factor for glaucomatous field loss. The higher the IOP, the greater the risk of optic nerve injury and loss of vision field. Brimonidine tartrate reduces intraocular pressure while having a negligible effect on cardiovascular and pulmonary markers.

Despite the fact that topical brimonidine (50–500 mg) caused miosis in certain animal experiments, there were no clinically relevant alterations in the size of the pupil in people. According to animal and human studies, Brimonidine has no substantial effect on the flow of blood to and from the eyes [63].

Changes in systolic and diastolic blood pressure, as well as changes in heart rate, were observed in patients and healthy volunteers treated with therapeutic doses of topical Brimonidine. However, these changes were not related to any adverse clinical consequences. In three large comparison studies, brimonidine 0.2% twice daily for up to 12 months had a minor effect on systolic (mean change from 3.52 to +0.64 mmHg) and diastolic blood pressure (mean change from 1.7 to +1.04 mmHg) and heart rate (0.1–3.1 beats/min) [63].

6.3 Pharmacokinetics [63,64]
6.3.1 Absorption

In a pharmacokinetic study, four males and 10 females received a single topical ocular dosage of Brimonidine Tartrate Ophthalmic Solution (0.15%)

one drop per eye in a pharmacokinetic study. The maximum plasma concentrations (C_{max}) and area under the curve (AUC_{0-inf}) were 73 ± 19 and
375 ± 89 pg h/mL, respectively. After dosing, the T_{max} was 1.7 ± 0.7 h.
The half-life in the system was approximately 2.1 h [63,65].

6.3.2 Metabolism

Brimonidine is metabolized by the liver in the majority of cases. According
to in vitro human liver microsomal and liver slice metabolic data,
brimonidine is extensively metabolized in the liver [63,65].

6.3.3 Excretion

Brimonidine and its metabolites are primarily eliminated through the kidneys, which is the most common route of elimination. Brimonidine was
excreted from the body in roughly 87% of cases after being administered
orally. After being delivered orally, approximately 74% of the radioactivity
was recovered in the urine [63,65].

6.4 Dosage and administration

The recommended dose of brimonidine for the treatment of open-angle
glaucoma and ocular hypertension is one drop of 2% solution put into
the afflicted eye (s). In the United States, the recommended dosing regimen
is three times daily, whereas it is twice daily in all other countries where the
medicine has been approved for use [63].

Brimonidine has the potential to damage one's ability to drive or operate
heavy machinery. Patients with severe cardiovascular illness, hepatic or renal
impairment, depression, cerebral or coronary insufficiency, Raynaud's phe-
nomenon, orthostatic hypotension, or thromboangiitis obliterans should be
advised to use caution when taking this medication. It is not recommended
that brimonidine be administered to patients who are taking monoamine
oxidase inhibitors. Patients using concomitant adrenoceptor antagonists,
antihypertensives, cardiac glycosides, or tricyclic antidepressants should be
treated with caution as well [63].

People who wear soft contact lenses should wait at least 15 min after taking their medicine before putting their lenses in Ref. [63].

Adults of all ages: Use one drop in each affected eye twice a day, 12 h apart.
In the elderly, no dosage change is necessary. One minute after each drop is
injected, the lachrymal sac should be constricted at the medial canthus (pun-
ctal occlusion) to limit the risk of systemic absorption. When more than one
topical ophthalmic medication is used, instill them at least 5 min apart.

Topical Gel: It is indicated in adult patients for the symptomatic treatment of rosacea facial erythema.

Adults and the elderly: Apply brimonidine in a thin coating throughout the entire face (front, chin, nose, and both cheeks), avoiding the inner nose's eyes, eyelids, lips, mouth, and membrane. Brimonidine should be used solely on the face. Hands should be cleansed shortly following the application of the medication. Brimonidine may be used in concert with other cutaneous medical agents to treat inflammatory lesions associated with rosacea and cosmetics. These products should not be used immediately prior to the daily administration of Brimonidine; they should be used only after the Brimonidine has dried [66].

Acknowledgments

We would like to thank ChemAxon for making the MARVIN software available as part of their free Academic Package.

References

[1] National Center for Biotechnology Information. PubChem Compound Summary for CID 3081361, Vandetanib. Available from: https://pubchem.ncbi.nlm.nih.gov/compound/Vandetanib. Accessed Jan. 29, 2022.

[2] Vandetanib Theoretical Analysis. Available from: https://www.medkoo.com/products/4783 Accessed Feb. 6, 2022.

[3] L.A. Anderson, S. Sinha, K. Durbin, S. Entringer, J. Stewart, P. Thornton, C. Fookes, M. Puckey, N. France, J. Grigg, S. Chao, Brimonidine, 2021. Available from: https://www.drugs.com/pro/brimonidine.html#moreResources.

[4] J.O.N. Maryadele (Ed.), The Merck Index. An Encyclopedia of Chemicals, Drugs and Biologicals, fourteenth ed., Merck & Co., Inc., Whitehouse Station, NJ, 2006, p. 683.

[5] https://en.wikipedia.org/wiki/Brimonidine. 12/09/2021.

[6] *Caprelsa Assessment report by EMA*. Available from: https://www.ema.europa.eu/en/documents/assessment-report/caprelsa-epar-public-assessment-report_en.pdf Accessed Feb. 11, 2022.

[7] S. Kharidia, USP-NF brimonidine tartrate, in: The United States Pharmacopeia (USP), USP 42 and NF 37, United States Pharmacopeial Convention, 2020, p. 578.

[8] Marvin 21.19.0-13698. ChemAxon. 2021.

[9] https://www.tradeindia.com/products/brimonidine-tartrate-79570-19-7-5247025.html. 12/12/2021.

[10] B.P. Bandgar, S.S. Sawant, S.M.S.J. Mukarram, A process for preparing brimonidine or salts thereof, Quickcompany, 2013 India. 27 pp. (https://www.quickcompany.in/patents/a-process-for-preparing-brimonidine-or-salts-thereof#documents).

[11] J.C. Danilewicz, M. Snarey, G.N. Thomas, (Imidazolinylamino) Quinazolines, -Quinolines, and -Quinoxalines, Pfizer Corp, 1973. 25 pp.

[12] M.B. Thomas, M.T. Williams, Quinoxaline and Quinazoline Derivatives, Pfizer Inc., Panama, 1976. 12 pp.

[13] J.C. Danielewicz, M. Snarey, G.N. Thomas, in: P. Inc (Ed.), (2-Imidazolin-2-ylamino) substituted-quinoxalines and-quinazolines as antihypertensive agents, June 14, United States Patent, New York, N.Y, 1977, pp. 1–10.

[14] Q. Chen, D. Chen, C. Wu, Preparation of 5-Bromo-6-Aminoquinoxaline as Intermediate of Brimonidine Tartrate, Jiaxing Institute of Applied Chemistry and Engineering, Chinese Academy of Sciences, Peop. Rep. China, 2008. 7 pp.

[15] L. Zhang, Q. Chen, C. Wu, Z. Zhang, D. Chang, Mild Synthesis of 6-Amino-5-Bromoquinoxaline, vol. 29, Xiandai Huagong, 2009, pp. 54–56. (Copyright (C) 2021 American Chemical Society (ACS). All Rights Reserved.):.

[16] L. Zhang, Q. Chen, C. Wu, D. Chang, C. Ha, Bromination Synthesis of 6-Amino-5-Bromoquinoxaline, vol. 26, Jingxi Huagong, 2009, pp. 575–579 (Copyright (C) 2021 American Chemical Society (ACS). All Rights Reserved).

[17] R.G. Glushkov, A.I. L'Vov, L.N. Dronova, G.A. Modnikova, I.E. Mamaeva, Improved Process for Preparation of 5-Bromo-6-[(2-Imidazolin-2-yl)Amino] Quinoxaline L-Tartrate as Antiglaucoma Agent, 2006. FGUP "Tsentr po Khimii Lekarstvennykh Sredstv", Russia. 4 pp.

[18] H.G. Viehe, Z. Janousek, The chemistry of dichloromethylenammonium salts ("phos-genimonium salts"), Angew Chem Int.Ed 12 (10) (1973) 806–818.

[19] A.M. Naik, S.D. Sawant, G.A. Kavishwar, S.G. Kavishwar, Novel process for the syn-thesis of Brimonidine and derivative, Int. J. PharmTech Res. 2 (2010) 14–17. (Copyright (C) 2021 American Chemical Society (ACS). All Rights Reserved.):.

[20] Attwa, M. W., Kadi, A. A., Darwish, H. W., Amer, S. M., & Al-Shakliah, N. S. (2018). Identification and characterization of in vivo, in vitro and reactive metabolites of vandetanib using LC–ESI–MS/MS. Chem. Cent. J. 12 (1), 99.

[21] M. Rawlins (Ed.), Brimonidine Tartrate, British Pharmacopoeia Commission, British Pharmacopoeia, vol. 1, The Stationery Office: Healthcare products Regulatory Agency (MHRA)., 2016, p. I-315.

[22] S.X. Xiang, H.-L. Wu, C. Kang, L.X. Xie, X.-L. Yin, H.-W. Gu, R.Q. Yu, Fast quan-titative analysis of four tyrosine kinase inhibitors in different human plasma samples using three-way calibration-assisted liquid chromatography with diode array detection, J. Sep. Sci. 38 (16) (2015) 2781–2788.

[23] Human Metabolome Database (HMDB) (Brimonidine), 2021. cited 2021; Available from: https://hmdb.ca/metabolites/HMDB0014627.

[24] S. Jiang, A.K. Chappa, J.W. Proksch, A rapid and sensitive LC/MS/MS assay for the quantitation of brimonidine in ocular fluids and tissues, J. Chromatogr. B 877 (3) (2009) 107–114.

[25] V. Radulovic, M.M. Aleksic, D. Agbaba, V. Kapetanovic, An electroanalytical approach to brimonidine at boron doped diamond electrode based on its extensive voltammetric study, Electroanalysis 25 (2013) 230–236. (Copyright (C) 2021 American Chemical Society (ACS). All Rights Reserved.).

[26] M.M. Aleksić, V. Radulović, D. Agbaba, V. Kapetanović, An extensive study of elec-trochemical behavior of brimonidine and its determination at glassy carbon electrode, Electrochim. Acta 106 (2013) 75–81.

[27] V. Radulović, M. Aleksić, V. Kapetanović, K.K. Rajić, M. Jovanović, I. Marjanović, M. Stojković, D. Agbaba, The evaluation of short-and long-term stability studies for brimonidine in aqueous humor by DPV/BDDE method—possible application for direct assay in native samples, Anal. Bioanal. Chem. 411 (22) (2019) 5755–5763.

[28] A.M. Abou Al Alamein, H.A. Hendawy, N.O. Elabd, Chemometrics-assisted voltammetric determination of timolol maleate and brimonidine tartrate utilizing a car-bon paste electrode modified with iron (III) oxide nanoparticles, Microchem. J. 145 (2019) 313–329.

[29] M.M. Annapurna, Multi component mode and derivative spectrophotometric methods for the simultaneous determination of timolol maleate and brimonidine tartrate, Asian J. Pharm. 12 (01) (2018) 251–255.

[30] M.C. Damle, A. Zirange, Derivative spectrophotometric method for estimation of brimonidine tartrate & timolol maleate in combination, World J. Pharm. Res. 7 (12) (2018) 725–731.

[31] M. Annapurna, M. Sushmitha, V. Sevyatha, A. Narendra, New spectrophotometric methods for the simultaneous determination of brimonidine and timolol in eye drops, JCHPS 10 (2) (2017) 786–789.

[32] M.M. Annapurna, R. Narendra, M.R. Pattanaik, Novel analytical spectrophotometric methods for the quantification of brimonidine tartrate—an α adrenergic agonist, J. Chem. Pharm. Sci. 8 (2015) 509–514. (Copyright (C) 2021 American Chemical Society (ACS). All Rights Reserved.):.

[33] J.F. de Souza, K.N. Maia, S.L. Fialho, F.P. de Andrade, G.R. da Silva, Development and validation of spectrophotometric method for the determination of brimonidine in ocular implants, World J. Pharm. Pharm. Sci. 3 (2014) 927–941 (Copyright (C) 2021 American Chemical Society (ACS). All Rights Reserved.).

[34] H.S. Popaniya, H.M. Patel, Simultaneous determination of brimonidine tartrate and timolol maleate in combined pharmacetical dosage form using two different green spectrophotometric methods, World J. Pharm. Pharm. Sci. 3 (2014) 1330–1340. (Copyright (C) 2021 American Chemical Society (ACS). All Rights Reserved.). 11.

[35] M.S. Rizk, H.A. Merey, S.M. Tawakkol, M.N. Sweilam, Simultaneous determination of Timolol maleate and Brimonidine tartarate in their pharmaceutical dosage form, Anal. Chem. Lett. 4 (2) (2014) 132–145.

[36] P.S. Jain, R.N. Khatal, S.J. Surana, Development and validation of first order derivative UV-spectrophotometric method for determination of brimonidine tartrate in bulk and in formulation, Asian J. Pharm. Biol. Res. 1 (3) (2011) 323–329.

[37] E.S. Elzanfaly, A.S. Saad, B. Abd Elaziz, A smart simple spectrophotometric method for simultaneous determination of binary mixtures, J. Pharm. Anal. 2 (5) (2012) 382–385.

[38] V.L. Jadhav, A.S. Patil, S.R. Chaudhari, A.A. Shirkhedkar, UV-Spectrophotometry-multicomponent mode of analysis for simultaneous estimation of brinzolamide and brimonidine tartrate in bulk and ophthalmic formulation, J. Pharm. Technol. Res. Manag. 7 (1) (2019) 31–35.

[39] F. Ibrahim, N. El-Enany, R.N. El-Shaheny, I.E. Mikhail, Validated spectrofluorimetric and spectrophotometric methods for the determination of brimonidine tartrate in ophthalmic solutions via derivatization with NBD-Cl. Application to stability study, Luminescence 30 (2015) 309–317. (Copyright (C) 2021 American Chemical Society (ACS). All Rights Reserved.):.

[40] G. Sunitha, R. Bhagirath, V.R. Alapati, K. Ramakrishna, C.V.S. Subrahmanyam, P.D. Anumolu, Fluorimetric quantification of brimonidine tartrate in eye drops, Indian J. Pharm. Sci. 75 (2013) 730–732. (Copyright (C) 2021 American Chemical Society (ACS). All Rights Reserved.):.

[41] D.N. Tzovolou, F. Lamari, E.K. Mela, S.P. Gartaganis, N.K. Karamanos, Capillary electrophoretic analysis of brimonidine in aqueous humor of the eye and blood sera and relation of its levels with intraocular pressure, Biomed. Chromatogr. 14 (2000) 301–305. (Copyright (C) 2021 American Chemical Society (ACS). All Rights Reserved.).

[42] S. Mandan, A. Nerpagar, U. Laddha, Development and validation of HPLC method for estimation of brimonidine tartrate as bulk drug and in ophthalmic formulation, J. Drug Deliv. Ther. 7 (7) (2017) 146–159.

[43] M.M. Baker, T.S. Belal, Validated HPLC–DAD method for the simultaneous determination of six selected drugs used in the treatment of glaucoma, J. AOAC Int. 101 (4) (2018) 993–1000.

[44] V.P. Agrawal, S.S. Desai, G.K. Jani, Development of RP-HPLC method for simultaneous determination of brimonidine tartrate and brinzolamide by QbD approach and its
validation, Eurasian J. Anal. Chem. 11 (2) (2017) 63–78.

[45] J. Christian, K.G. Patel, Validation and experimental design assisted robustness testing of
RPLC method for the simultaneous analysis of brinzolamide and brimonidine Tartrate
in an ophthalmic dosage form, Indian J. Pharm. Sci. 78 (5) (2016) 631–640.

[46] M. Walash, R. El-Shaheny, Fast separation and quantification of three anti-glaucoma
drugs by high-performance liquid chromatography UV detection, J. Food Drug
Anal. 24 (2) (2016) 441–449.

[47] B. Rohan, T. Amol, K. Chandrakant, B. Shrivastava, Development and validation of
high performance liquid chromatographic method for estimation of brimonidine tartrate as bulk drug and in ophthalmic formulation, Int. J. PharmTech Res. 12
(2019) 99–105 (Copyright (C) 2021 American Chemical Society (ACS). All Rights
Reserved.).

[48] K.N. Rao, R. Doonaboyina, R. Hema, Method development and validation of
brinzolamide and brimonidine in its bulk and ophthalmic dosage form by using
RP-HPLC, Int. J. Chem. Pharm. Sci. (Nellore, India) 6 (2018) 306–312 (Copyright
(C) 2021 American Chemical Society (ACS). All Rights Reserved.).

[49] F.A. Ibrahim, H.M. Elmansi, S.A. El Abass, A versatile HPLC method with an isocratic
single mobile phase system for simultaneous determination of anti-glaucoma formulations containing timolol, Ann. Pharm. Fr. 77 (2019) 302–312. (Copyright (C) 2021
American Chemical Society (ACS). All Rights Reserved.).

[50] U.D. Laddha, S.S. Mandan, A.V. Nerpagar, S.J. Surana, Development and validation of
reverse phase high performance liquid chromatographic method for estimation of
brimonidine tartrate, Int. J. Pharma Res. Rev. 5 (2016) 1–5. (Copyright (C) 2021
American Chemical Society (ACS). All Rights Reserved.).

[51] V.P. Agrawal, S.S. Desai, G.K. Jani, Development of RP-HPLC method for simultaneous determination of brimonidine tartrate and brinzolamide by QbD approach and its
validation, Eurasian J. Anal. Chem. vol. 11 (2016) 63–78 (Copyright (C) 2021
American Chemical Society (ACS). All Rights Reserved.).

[52] J.R. Christian, K. Patel, T.R. Gandhi, Validation and experimental design assisted
robustness testing of rplc method for the simultaneous analysis of brinzolamide and
brimonidine tartrate in an ophthalmic dosage form, Indian J. Pharm. Sci. 78
(2016) 631–640 (Copyright (C) 2021 American Chemical Society (ACS). All Rights
Reserved.).

[53] J. Sun, X. Zhang, T. Huang, A validated stability-indicating HPLC method for determination of brimonidine tartrate in BRI/PHEMA drug delivery systems, Chem. Cent.
J. 11 (1) (2017) 1–10.

[54] A.A. Elshanawane, L.M. Abdelaziz, M.S. Mohram, H.M. Hafez, Development and validation of HPLC method for simultaneous estimation of brimonidine tartrate and timolol maleate in bulk and pharmaceutical dosage form, J. Chromatogr. Sep. Tech. 5
(3) (2014) 1.

[55] F. Ibrahim, H. Elmansi, S. El Abass, A versatile HPLC method with an isocratic single
mobile phase system for simultaneous determination of anti-glaucoma formulations
containing timolol, in: Annales pharmaceutiques francaises, Elsevier, 2019.

[56] P. Patel, V.C. Darji, B.R. Patel, Analytical method development and validation of stability indicating RP-HPLC method for estimation of brinzolamide and brimonidine
tartrate in an ophthalmic suspension, IJRAR 6 (2019) 150–156.

[57] H.M. Hafez, A.A. Elshanawane, L.M. Abdelaziz, M.S. Mohram, Design of experiment
utilization to develop a simple and robust RP-UPLC method for stability indicating
method of anti glaucoma ophthalmic drops, Eurasian J. Anal. Chem. 10 (1) (2015)
46–67.

[58] A. Narendra, D. Deepika, M.M. Annapurna, Liquid chromatographic method for the analysis of brimonidine in ophthalmic formulations, E-J. Chem. 9 (3) (2012) 1327–1331.

[59] S.T. Hassib, E.F. Elkady, R.M. Sayed, Simultaneous determination of timolol maleate in combination with some other anti-glaucoma drugs in rabbit aqueous humor by high performance liquid chromatography-tandem mass spectroscopy, J. Chromatogr. B Analyt. Technol. Biomed. Life Sci. 1022 (2016) 109–117 (Copyright (C) 2021 American Chemical Society (ACS). All Rights Reserved.).

[60] Z. Tang, X. Li, H. Xu, S. Chen, B. Wang, Q. Wang, HPLC–MS/MS studies of brimonidine in rabbit aqueous humor by microdialysis, Bioanalysis 13 (19) (2021) 1487–1499.

[61] V. Radulovic, M. Aleksic, V. Kapetanovic, K.K. Rajic, M. Jovanovic, I. Marjanovic, M. Stojkovic, D. Agbaba, The evaluation of short- and long-term stability studies for brimonidine in aqueous humor by DPV/BDDE method-possible application for direct assay in native samples, Anal. Bioanal. Chem. 411 (2019) 5755–5763 (Copyright (C) 2021 American Chemical Society (ACS). All Rights Reserved.).

[62] L. Fala, Mirvaso (Brimonidine): First Topical Gel Approved by the FDA for the Treatment of Facial Erythema of Rosacea, American Health & Drug Benefits, 2021.

[63] J.C. Adkins, J.A. Balfour, Brimonidine, Drugs Aging 12 (3) (1998) 225–241.

[64] B. Limone, J. Luke, Brimonidine, J. Dermatol. Nurses Assoc. 10 (3) (2018) 154–157.

[65] ALPHAGAN, Highlights of Prescribing Information, [cited 2022 16/05]; Available from: https://www.accessdata.fda.gov/drugsatfda_docs/label/2016/020613s031lbl.pdf.

[66] Datapharm, 2021. Available from: https://www.medicines.org.uk/emc/product/5303/smpc.

CHAPTER TWO

Crizotinib: A comprehensive profile

Ahmed A. Abdelgalil[a] and Hamad M. Alkahtani[b]

[a]Central Laboratory, College of Pharmacy, King Saud University, Riyadh, Kingdom of Saudi Arabia
[b]Department of Pharmaceutical Chemistry, College of Pharmacy, King Saud University, Riyadh, Kingdom of Saudi Arabia

Contents

Profiles of Drug Substances, Excipients, and Related Methodology, Volume 48
ISSN 1871-5125
https://doi.org/10.1016/bs.podrm.2022.11.002

39

1. Physical profiles of drug substances and excipients

1.1 General information

1.1.1 Nomenclature

1.1.1.1 Systematic chemical names

3-[(1*R*)-1-(2,6-dichloro-3-fluorophenyl)ethoxy]-5-(1-piperidin-4-ylpyrazol-4-yl)pyridin-2-amine [1]

1.1.1.2 Nonproprietary names

Crizotinib, 77399-52-5, PF-02341066, (*R*)-crizotinib.

1.1.1.3 Proprietary names

Xalkori® by Pfizer.

1.2 Formulae

1.2.1 Empirical formula, molecular weight, CAS number [1]

Empirical Formula: $C_{21}H_{22}Cl_2FN_5O$

Molecular Weight: 450.337 g/mol

CAS Number: 877399-52-5

1.2.2 Structural formula

See Fig. 1

1.2.3 Simplified molecular input line entry system (SMILES)

C[C@@H](OC1=CC(=CN=C1N)C1=CN(N=C1)C1CCNCC1)C1=C(Cl)C=CC(F)=C1Cl [1]

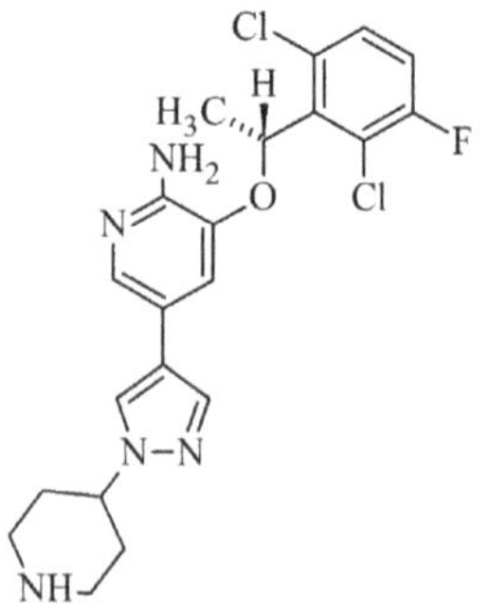

Fig 1 Crizotinib structure.

1.2.4 The IUPAC International Chemical Identifier (InChI)

InChI=1S/C21H22Cl2FN5O/c1-12(19-16(22)2-3-17(24)20(19)23)30-
18-8-13(9-27-21(18)25)14-10-28-29(11-14)15-4-6-26-7-5-15/h2-3,8-12,
15,26H,4-7H2,1H3,(H2,25,27)/t12-/m1/s1 [1]

1.3 Elemental analysis

Carbon: 55.46%
Hydrogen: 4.95%
Nitrogen: 15.60%

1.4 Appearance

Crizotinib is a white- to pale-yellow powder [2]

2. Physical characteristics

2.1 Ionization constants

$pK_a = 10.12$ [1].

2.2 Solubility characteristics

Soluble in DMSO at 25 mg/mL with warming; soluble in ethanol at 25 mg/mL with warming [3]. Very poorly soluble in water; water solubility: 0.00611 mg/mL with maximum solubility in plain water 10–20 μM [1]. The solubility of crizotinib in aqueous media decreases over the range of pH 1.6–8.2 from >10 mg/mL to <0.1 mg/mL [2].

2.3 Partition coefficients

The octanol/water partition coefficient (LogP) value reported for crizotinib is 1.83 [4].

2.4 Crystallographic properties

2.4.1 X-ray powder diffraction pattern

Powder X-ray diffraction of the Crizotinib (free base) was evaluated by Ultima IV diffractometer (Rigaku, Japan) over the 3–140 then 3−60° 2θ range at a scan speed of 1 degree and 0.5 degree/min. The tube anode was Cu with Ka=0.1540562 nm monochromatized with a graphite crystal. The pattern was collected at 40 kV of tube voltage and 40 mA of tube current in step scan mode (step size 0.02°, counting time 1 s per step). The X-ray powder diffraction pattern of crizotinib is given in the Fig. 2, Table 1. The powder X-ray diffraction pattern showed peaks at diffraction angles (2θ) of 17.3 + 0.1 and 19.7 + 0.1.

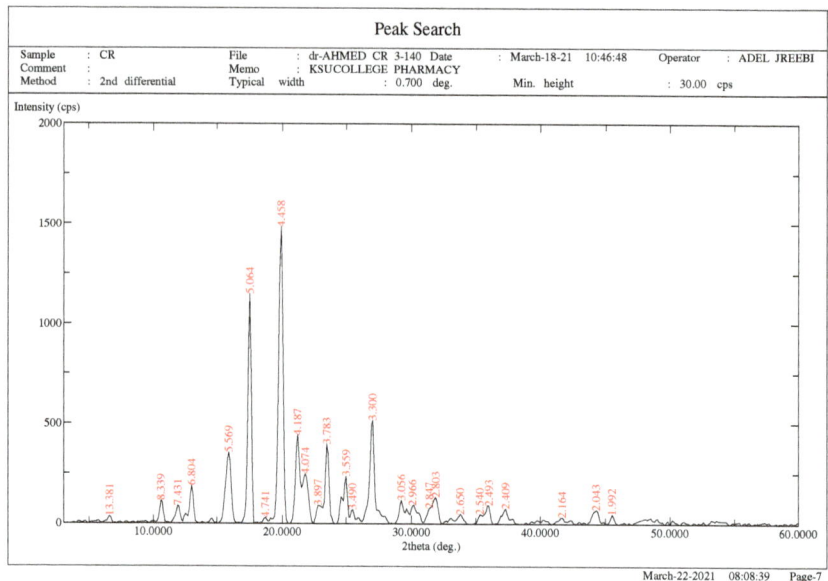

Fig 2 X-ray powder diffraction pattern of Crizotinib.

Table 1 XRD analysis of crizotinib powder (free base).

Angel (2theta)	D-value(Å)	Intensity (%)
6.600	13.3813	34
10.600	8.3390	113
11.900	7.4308	89
13.000	6.8044	189
15.900	5.5693	355
17.500	5.0635	**1149**
18.700	4.7412	30
19.900	4.4579	**1483**
21.200	4.1874	442
21.800	4.0735	248
22.800	3.8971	90
23.500	3.7825	397
25.000	3.5589	237

Table 2 ^{1}H NMR of crizotinib.

Signal	Chemical shift	Splitting	Integration	Assignment
3 and 4	1.75	qd (J = 3.54 and 12.02 Hz)	2	C**H** (axial)
–	1.80	d (J = 6.58 Hz)	3	C**H**$_3$
3 and 4	1.90–1.97	m	2	C**H** (equatorial)
1 and 2	2.57	t (J = 12.17 Hz)	3	C**H** (axial)
1 and 2	3.02	d (J = 12.28 Hz)	2	C**H** (equatorial)
5	4.14	tt (J = 4.0. and 11.60 Hz)	1	C**H**
–	5.64	br s	2	N**H**$_2$
10	6.08	q (J = 6.58 Hz)	1	C**H**
9	6.89	s	1	Pyridine
12	7.43	t (J = 8.61 Hz)	1	Ph**H**
6	7.52	s	1	Pyrazole
11	7.57	dd (J = 4.76 and 8.61 Hz)	1	Ph**H**
7	7.75	s	1	Pyrazole
8	7.91	s	1	Pyridine

2.6.3.2 ^{13}C NMR spectrum

The ^{13}C NMR spectra of crizotinib (Shown in Fig. 7) was scanned in deuterated DMSO on Bruker 700 MHz NMR spectrometer. The parameters derived from the various NMR spectra are presented in Table 3.

2.6.4 Mass spectrometry

The mass spectrum of crizotinib was obtained using Varian CP 3800 GC combined with 320 TQMS mass spectrometer as direct probe injection using high-purity helium as the gas carrier, at a flow rate of 1 mL/min. The source temperature of MS was set at 250 °C and the Quad temperature was at 250 °C The direct probe method initially at 90 °C (held for 2 min),

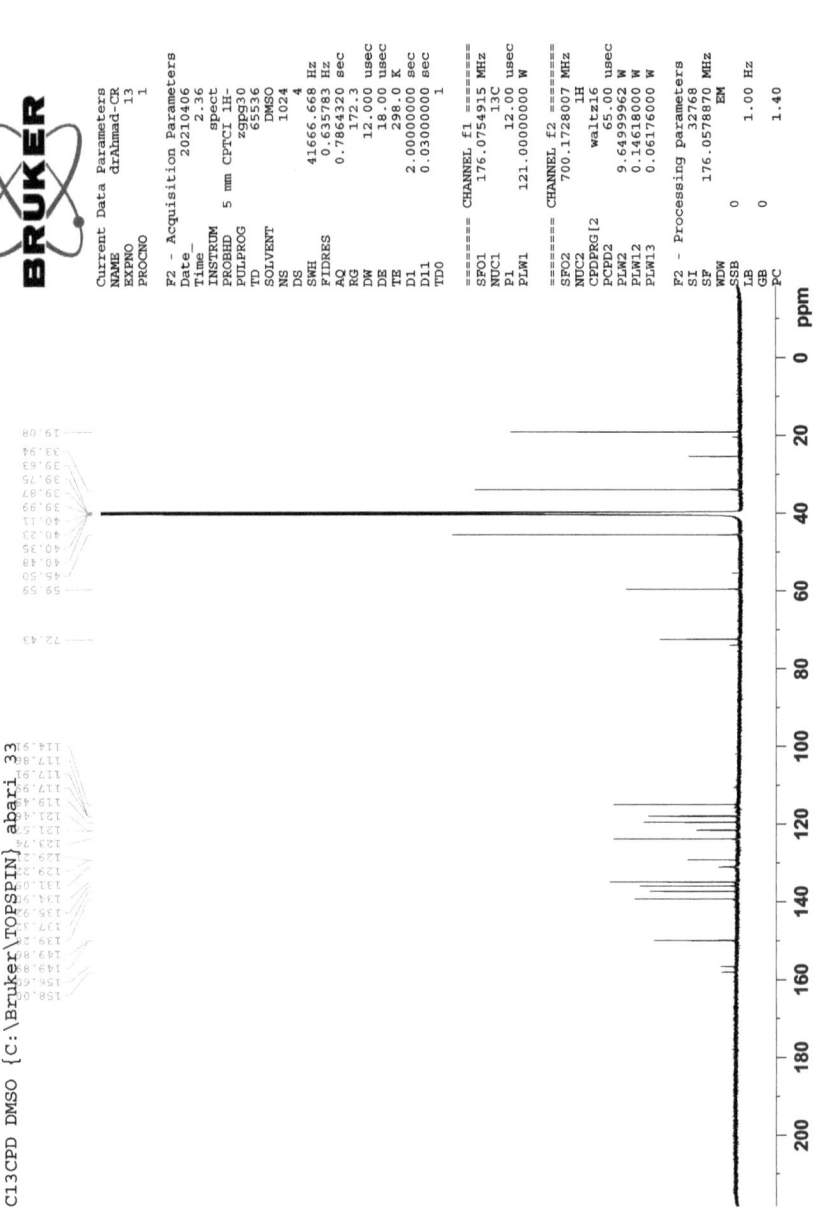

Fig 7 ^{13}C NMR spectrum.

Table 3 ^{13}C NMR of crizotinib.

Signal	Chemical shift	Assignment
–	18.61	C$\underline{\text{H}}_3$
1 and 2	33.47	$\underline{\text{C}}$H$_2$ (piperidine)
3 and 4	45.03	$\underline{\text{C}}$H$_2$ (piperidine)
5	59.12	$\underline{\text{C}}$H (piperidine)
14	71.96	—$\underline{\text{C}}$HOAr
7	114.44	Pyrazole
8	117.44	Pyrazole
18	117.46 ($J = 23.09$ Hz)	Ph
11	119.01	Pyridine
16	121.05 ($J = 19.23$ Hz)	Ph
19	123.27	Ph
20	128.75 ($J = 3.3$ Hz)	Ph
6	130.58	Pyrazole
9	134.43	Pyridine
15	135.45	Ph
12	136.84	Pyridine
10	138.79	Pyridine
13	149.41	Pyridine
17	156.90 ($J = 25.29$ Hz)	Ph

then was increased to 150 °C at 30 °C min^{-1} (held for 2 min), then increased further to 300 °C at 30 °C min^{-1} for 2 min. The capillary probe volume was 1 μL approx. and the scan range was set at 50–1000 mass ranges at 70 eV electron energy with no solvent delay. Fig. 8 shows the fragmentation pattern of crizotinib. The spectrum was identified with the help of library embedded in the instrument software.

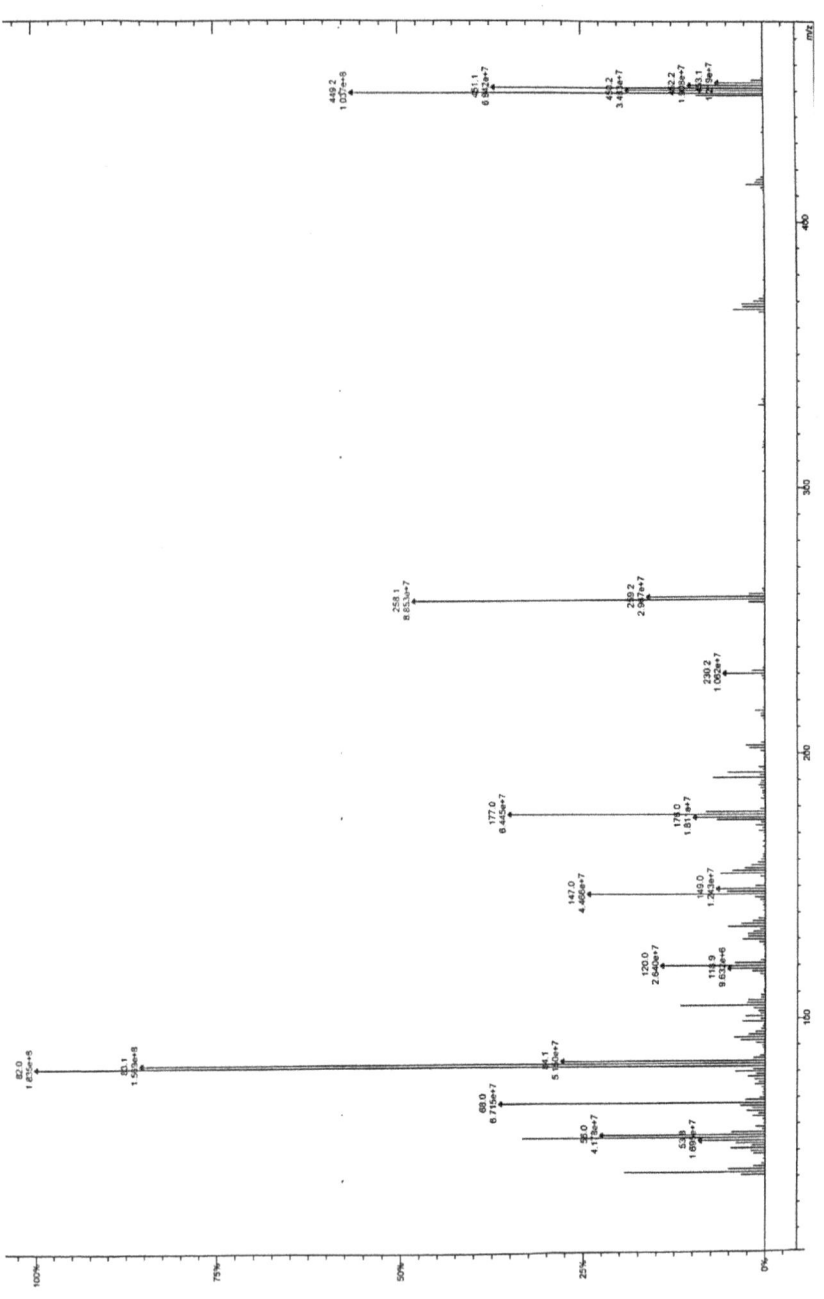

Fig 8 Fragmentation pattern of Crizotinib.

3. Stability

3.1 Solid-state stability

In the crizotinib package, insert it is recommended to store crizotinib capsules at room temperature 20–25 °C (68–77°F); excursions permitted between 15 and 30 °C (59 and 86°F) [6].

4. Analytical profiles of drug substances and excipients

4.1 Spectroscopic methods of analysis

4.1.1 Spectrophotometry

In series of studies, the reaction of 1,4-benzoquinone (BQ), chloranilic acid (CA) or 2,3-dichloro-1,4-naphthoquinone (DCNQ) with crizotinib (CZT) were studied in different solvents of varying dielec. constants and polarity indexes. The reactions gave a red, a violet or red-colored products. Authors reported that, spectrophotometric studies confirmed that the reaction proceeded through charge-transfer (CT) complex formation. The molar absorptivity of the complex is linearly correlated with the dielec. constant and polarity index of the solvent; the correlation coefficients were BQ-CZT (0.9425 and 0.8340, respectively) and DCNQ-CZT (0.9567 and 0.9069, respectively). The stoichiometric ratio of BQ:CZT was 2:1, and DCNQ:CZT was 2:1 and the association constant of the complex is $0.26 \times 10^3 \, \mathrm{L\,mol^{-1}}$, and $1.07 \times 10^2 \, \mathrm{L/mol}$ for DCNQ:CZT. The reaction was employed as a basis in the development of a novel 96-microwell assay for CZT. The assay limits of detections and quantitation were BQ-CZT 5.2 and $15.6 \, \mu\mathrm{g\,mL^{-1}}$, respectively, CA-CZT 8.8 and $26.4 \, \mu\mathrm{g\,mL^{-1}}$, respectively and 2.06 and 6.23 μg/mL, respectively for DCNQ:CZT [7–9].

Wani and Darwish described a 96-microwell-based high throughput spectrophotometric assay for pharmaceutical quality control of crizotinib (CZT). They examined the reaction between CZT and 1,2-naphthoquinone-4-sulphonate, a chromogenic reagent. A red-colored product showing a maximum absorption peak (λ_{max}) at 490 nm was produced in an alkaline medium (pH 9). Beer's law, which shows a correlation between absorbance and CZT concentration, was obeyed in the range of 4–50 μg/mL with an appropriate correlation coefficient (0.999). The limits of detection and quantification were 1.73 and 5.23 μg/mL, respectively [10].

4.1.2 Spectrofluorometric method

Darwish et al., developed a procedure to quantify the crizotinib, in human plasma and bulk powder by highly sensitive micellar enhanced spectrofluorimetric procedure. The method based on measuring the fluorescence intensity of crizotinib (CRZ) in sodium dodecyl sulphate (SDS) micellar system at 404 nm after excitation at 271 nm. Maximum fluorescence intensity (FI) was attained by addition of 0.2 mL SDS and 0.2 mL HCl (1N) to CRZ aliquots and then dilution with distilled water. There was a linear relationship between the FI of CRZ and its concentration over the range, 5–400 ng/mL, with limit of detection and of quantification of 1.857 and 5.628 ng/mL, respectively. The developed procedure was applied to assay CRZ in pure powder form and spiked human plasma with mean recovery of 100.68% ± 0.37% and 99.98% ± 0.20%, respectively [11].

Binding of crizotinib (CRB) with the bovine serum albumin (BSA) CRB was capable of linearly quench the BSA fluorescence. This quenching effect was studied over the emission wavelength range of 290–500 nm after being excited at 280 nm. CRB-BSA binding did not induce any changes either in the emission wavelength or peak shape of the BSA [12].

4.2 Enzyme-immunoassay methods

Al-Shehri et al. described an ELISA method for monitoring crizotinib plasma concentration. The method involved synthesis of hapten (crizotinib-acetohydrazide), which then coupled to {bovine serum albumin (BSA) and keyhole limpet hemocyanin (KLH) proteins by ethyl-3-(3-dimethylaminopropyl) carbodiimide as a coupling reagent). Polyclonal antibody was generated using Crizotinib-KLH conjugate (IC50 = 0.5 ng/mL). Then ELISA involved a competitive binding reaction between CZT, in its samples, and immobilized CZT-BSA conjugate for the binding sites on a limited amount of the anti-CZT antibody . The limit of detection and the working range were 0.03 ng/mL and 0.05–24 ng/mL, respectively. Analytical recovery of CZT from spiked plasma was 101.98% ± 2.99%. The precisions of the assay were satisfactory; RSD was 3.2–6.5% and 4.8–8.2%, for the intra- and inter-assay precision, respectively. The proposed ELISA is anticipated to effectively contribute to the therapeutic monitoring of CZT in clinical settings [13].

4.3 Chromatographic methods of analysis

4.3.1 High performance liquid chromatography

Crizotinib was quantified in different biological matrices such as (human plasma, serum, whole blood and tissues, cell culture and animal plasma) and in pharmaceutical dosage forms. Detailed description of reported HPLC and LC/MS methods for determination of crizotinib in different biological matrices and in pharmaceutical dosage forms. is shown in Table 4.

Table 4 Detailed description of reported HPLC methods for determination of crizotinib in different biological matrices and in pharmaceutical dosage forms.

No	Technique	Target	Matrix	Sample preparation	Mobile phase	Column	Detection	Linearity range	REF
1	LC-MS/MS	Crizotinib	Mouse plasma	Protein precipitation IS: Crizotinib-^{13}C2-^{2}H5	0.1% (v/v) ammonium hydroxide (A) and methanol (B); Gradient (0.6 mL/min)	Acquity UPLC® BEH C18 column (30 mm × 2.1 mm, dp = 1.7 μm) VanGuard pre-column (Waters, 5 mm × 2.1 mm	Selected reaction monitoring (SRM) mode	10–10,000 ng/mL with r2 = 0.99980 ± 0.00014 for double logarithmic linear regression (n = 5). Within day precisions (n = 6) were 3.4–4.8%, between day (3 days; n = 18) precisions 3.6–4.9%. Accuracies were between 107% and 112% for the whole calibration range	[4]
2	LC-ESI-MS/MS	Crizotinib	Human and mouse plasma	Solid-phase extraction (SPE)	Mobile phase A: water/formic acid 100:0.3, v/v) Mobile phase B: (MeOH/formic acid 100/0.3, v/v) 20% of B and 80% of A for 0.5 min then changed to 30% of B from 0.5 to 1 min and held until 4 min when the %B was increased to 85%. At 4.5 min the system was returned to starting conditions for a total sample run time of 7.0 min. Gradient elution	Supelco Discovery c18 column (50 mm × 2.1 mm, 5.0μ)	Multiple reaction monitoring + MRM m/z 450.2 > 260.2 for crizotinib and m/z 457.2 > 267.3 for the ISTD ([^{2}H5, ^{13}C2]-Crizotinib)	5–5,000 ng/mL for human plasma with correlation coefficients (r2) >0.998 and 2–2,000 ng/mL for mouse plasma with r2 values >0.999	[14]
3	LC-MS/MS	Crizotinib	Rat plasma	Simple protein precipitation with methanol–acetonitrile (1:1, v/v)	0.1% formic acid (A) and methanol containing 0.1% formic acid (B) together with the following gradient: 0.00 min 10% B, 0.30 min 10% B, 0.40 min 98% B, 1.80 min 98% B, 1.81 min 10% B, and 3.30 min 10% B, The flow rate was set at 0.50 mL/min. Gradient elution	Agilent Zorbax XDB C18 column (2.1 × 50 mm, 3.5 μm)	Multiple reaction monitoring MRM m/z = 450.3 → 177.1 for crizotinib and 386.2 → 122.2 for buspirone (IS)	1–2000 ng/mL The LLOQ was 1.00 ng/mL using 50 μL of rat plasma The mean recoveries of crizotinib in rat plasma were calculated to be 91.3%, 94.9% and 95.8% at three concentration levels at 5 replicates, with the RSD values of 1.93%, 2.53% and 0.51%, respectively	[15]

Continued

No	Technique	Target	Matrix	Sample preparation	Mobile phase	Column	Detection	Linearity range	REF
4	LC-MS/MS	Crizotinib	Mouse tissues	Protein precipitation with methanol	Methanol (solvent A) and 0.3% formic acid water solution (solvent B), and the flow rate was 0.3 mL/min. The linear gradient program was as follows: 0 to 0.2 min, 10% A; 0.2 to 0.5 min, 10 to 60% A; 0.5 to 1.9 min, 60% A; 1.9 to 2.1 min, 60 to 100% A; 2.1 to 3.1 min, 100% A; 3.1 to 3.5 min, 100 to 10% A; and 3.5 to 5.0 min, 10% A. Gradient elution	Phenomenex Kinetex C18 (50 mm × 2.1 mm, 2.6 μm)	Multiple reaction monitoring + MRM m/z 450.1 → 260.2 for crizotinib and m/z 398.2 → 212.0 for apatinib	20–8000 ng/mL (r > 0.99, n = 8)	[16]
5	LC-MS/MS	Gefitinib, erlotinib, icotinib, crizotinib, lapatinib and apatinib	Human plasma	Liquid–liquid extraction (ethyl acetate: *tert*-Butyl methyl ether, 1:1 v/v)	0.1% formic acid (A) and methanol (B) with a flow rate of 0.4 mL/min. The gradient elution program was 20% B from 0 to 0.5 min, 20–95% B from 0.5 to 3.0 min, 95% B from 3.0 to 6.0 min, 95–20% B from 6.0 to 6.1 min, and 20% B from 6.1–8.0 min. Gradient elution	Hypersil GOLD-C18column (50 × 2.1 mm, 5 μm, Thermo Scientific)	Multiple reaction monitoring + MRM (m/z): 398.1975/212.0818 for apatinib, 450.1258/260.1506 for crizotinib, 394.1761/336.1343 for erlotinib, 447.1593/128.1070 for gefitinib, 392.1604/304.1081 for icotinib, 581.1420/458.1066 for lapatinib and 494.2662/394.1662 for imatinib	3.0–2000 ng/mL for apatinib	[17]

	Technique	Analytes	Matrix	Extraction	Detection mode	Column	Mobile phase / elution	Validation	Ref.
6	LC-MS/MS	Afatinib, axitinib, ceritinib, crizotinib, dabrafenib, enzalutamide, regorafenib and trametinib	Human plasma	Protein precipitation with acetonitrile	Multiple reaction monitoring + MRM	Phenomenex Gemini C18 column (5.0 μm, 50 × 2.0 mm)	40% B, 250 μL/min (0-0.1 min); 40-100% B, 250 μL/min (0.1-2.0 min); 100% B, 250 μL/min (2.0-4.0 min); 100-40% B, 250 μL/min (4.0-4.01 min); 40% B, 500 μL/min (4.01-4.5 min); 40% B, 250 μL/min (4.5-5.0 min). Flow of 0.250 mL/min. Gradient elution	2.0-200 ng/mL for afatinib, axitinib, dabrafenib and trametinib and from 50.0 to 5000 ng/mL for ceritinib, crizotinib, enzalutamide and regorafenib were linear and correlation coefficients (r2) of 0.996 or better were obtained	[18]
7	LC-MS/MS	Afatinib, axitinib, bosutinib, crizotinib, dabrafenib, dasatinib, erlotinib, gefitinib, imatinib, lapatinib, nilotinib, ponatinib, regorafenib, regorafenib M2, regorafenib M5, ruxolitinib, sorafenib, sunitinib, vandetanib	Human plasma	Solid phase extraction	Multiple reaction monitoring + MRM	CORTECS® C18 UPLC dp = 1.6 μm, 2.1 × 50 mm	Mobile phase A was an acetic acid buffer 0.01% in water mobile phase B was acetonitrile added with 10% A. The following gradient was applied: starting at 12% B/88% A mobile phase, increasing linearly to 15% B at 0.25 min, 23% B at 0.75 min, 30% B at 1.5 min, 40% B at 2 min, 60% at 2.9 min, and 90% at 3.6 min. At 4.2 min, B was reduced to 12% and stayed at the same ratio until 5 min, the end of the run. Gradient elution	Drugs were arranged in four groups, according to their plasma concentration range: 0.1-200 ng/mL, 1-200 ng/mL, 4-800 ng/mL and 25-5000 ng/mL Correlation coefficients (r2) were at least 0.997 with a slope CV lower than 15% (n=3 for each molecule)	[19]
8	UPLC-MS/MS	Afatinib, crizotinib, osimertinib, erlotinib and nintedanib	Human plasma	protein precipitation with acetonitrile containing 1% (v/v) formic acid	Selected reaction monitoring mode (SRM)	Accucore® C18 (2.1 × 50 mm; 2.6 μm)	Mobile phase A: water acidified with 0.1% (v/v) formic acid, and Mobile phase B: acetonitrile containing 0.1% (v/v) formic acid at a flow rate of 500 μL/min. Gradient elution	Afatinib 5-250 ng/mL, Crizotinib 5-1000 ng/mL, osimertinib 5-1000 ng/mL, erlotinib 20-4000 ng/mL, nintedanib 1-250 ng/mL Intra- and inter-assay coefficient of variation from 2.6% to 10.6%), accuracy (from 96.1% to 108.5%), recovery and matrix effects	[20]

Continued

Table 4 Detailed description of reported HPLC methods for determination of crizotinib in different biological matrices and in pharmaceutical dosage forms.—cont'd

No	Technique	Target	Matrix	Sample preparation	Mobile phase	Column	Detection	Linearity range	REF
9	LC-MS/MS	Crizotinib and its major oxidative metabolite crizotinib-lactam	Human plasma	protein precipitation with acetonitrile containing 0.1% formic acid	0.1% formic acid aqueous (A) and acetonitrile/methanol (v: v, 1:1, B) as mobile phase, which was delivered at a flow rate of 0.3 mL/min. The gradient elution procedure was optimized as follow: 0–0.4 min 10%–25% B, 0.4–1.4 min 25%–50% B, 1.4–2.2 min 50%–90% B, 2.2–3.0 min 10% B. Gradient elution	ACQUITY UPLC BEH C18 column (50 × 2.1 mm, i. d., 1.7 μm)	positive selected reaction monitoring mode	0.1–1000 ng/mL for crizotinib and 0.1–400 ng/mL for crizotinib-lactam Correlation coefficients >0.999 (r > 0.999). The extraction recovery was >87.12%. The precision (RSD, %) was <8.27%, whereas accuracy (RE, %) was within the range of −4.56% to 7.08%	[21]
10	UPLC-MS/MS	Afatinib, alectinib, crizotinib and osimertinib	Human plasma	—	—	Acquity UPLC ® BEH C18; 2.1 × 50 mm, 1.7 μm	—	1.00–100 ng/mL for afatinib and 10.0–1000 ng/mL for alectinib, crizotinib and osimertinib lower limits of quantification at 1.00 ng/mL for afatinib and 10.0 ng/mL for alectinib, crizotinib and osimertinib. Within-run and between-run precision measurements fell within 10.2%, with accuracy ranging from 89.2% to 110%	[22]

No.	Technique	Analytes	Matrix	Sample preparation	Mobile phase	Column	Detection	Results	Ref.
12	HPLC-MS/MS	Alectinib, crizotinib, erlotinib and gefitinib	Human plasma	Protein precipitation with methanol	Ammonium acetate in water and in methanol, both acidified with formic acid 0.1%. Gradient elution	HyPURITY® C18	MRM, Transitions used for the different compounds were as follows: 483.3–396.2 (alectinib), 450.2–260.1 (crizotinib), 394.2–278.0 (erlotinib), 447.2–28.1 (gefitinib)	100–2000 ng/mL for alectinib and erlotinib and 50–1000 ng/mL for crizotinib and gefitinib. The coefficients of determination were 0.9906–0.9990 for alectinib, 0.9924–0.9993 for crizotinib, 0.9922–0.9998 for erlotinib and 0.9903–0.9999 for gefitinib	[23]
13	UPLC-MS/MS	Alectinib, ceritinib, and crizotinib	Wistar rat plasma	Protein precipitation with acetonitrile	Mixture of acetonitrile/water containing 0.1% formic acid (70: 30, v/v). Isocratic elution	Waters BEH™ C18 column	Multiple reaction monitoring (MRM)	Alectinib, ceritinib, and crizotinib concentration ranges; 2–200, 0.4–200, and 4.0–200ng/mL, respectively correlation coefficients of the three drugs, r ≥ 0.9997	[24]
14	LC-MS/MS	Afatinib, alectinib, ceritinib, crizotinib, dacomitinib, erlotinib, gefitinib, and osimertinib	Human serum	Liquid extraction method	—	—	Positive ionization mode	Calibration curves were linear across concentration ranges examined. The intra- and interassay accuracies were 90.7%–110.7% and 94.7%–107.6%, respectively. the mean recovery and percent coefficient of variation values ranged between 54%–112% and 1.7%–11.7%, respectively	[25]
16	RP-HPLC	Crizotinib	Capsule dosage form	—	Methanol and KH2PO4 buffer (pH 3). Isocratic elution	Syncronis (25 cm × 4.6mm)	UV 267 nm	10μg–1000μg/mL (R2 0.999)	[26]

Continued

Table 4 Detailed description of reported HPLC methods for determination of crizotinib in different biological matrices and in pharmaceutical dosage forms.—cont'd

No	Technique	Target	Matrix	Sample preparation	Mobile phase	Column	Detection	Linearity range	REF
17	LC-MS	Crizotinib	Rat plasma	Protein precipitation by acetonitrile–methanol (9:1, v/v) IS: midazolam	Acetonitrile–0.1% formic acid in water. Gradient elution	Zorbax SB-C18 (2.1 mm × 150 mm, 5 μm)	An electrospray ionization source was applied and operated in positive ion mode; selective ion monitoring (SIM) mode was used for quantification using target fragment ions m/z 450 for crizotinib and m/z 326 for the IS	8–500 ng/mL Mean recoveries of crizotinib in rat plasma were in the range of 76.6–89.7%. RSD of intra–day and inter–day precision were both <16%	[27]
18	HPLC	Crizotinib	Human plasma	Protein precipitation using methanol IS: Telmisartan	Mobile phase consisted of 1% acetic acid:acetonitrile (70:30 v/v). isocratic mode	μ-Bondapack CN column, (150 mm length × 3.9 mm i.d., 5 μm particle diameter)	Fluorescence detection at a wavelength of 264 nm for excitation and 382 nm for emission	Concentration range of 2–512 ng/mL. The % recovery values from spiked human plasma for the intra-day anal. were 98.37–103.01% with a mean value of 100.69% ± 7.26% whereas the values for the inter-day anal. were 98.37–102.44% with a mean value of 100.26% ± 7.06%	[28]
19	UPLC	Crizotinib	Stability indicating method for detection of crizotinib in pharmaceutical dosage form (Capsules)	—	0.1% Ortho–phosphoric acid and acetonitrile (45:55% v/v). isocratic mode	Hibra C18 (100 mm × 2.1 mm, 2μ) column	The detection was done at a wavelength of 327 nm	Concentration range of 37.5 μg/mL–225 μg/mL, with a correlation coefficient of 0.999. The method was validated according to the ICH guidelines. The developed method was found to be accurate and precise, with % recovery 99.9%–100.18% and % relative standard deviation 1.1	[29]

	Technique	Drug	Sample	Mobile phase	Column	Detection	Results	Ref
20	HPLC	Crizotinib	human plasma using di-Et ether as extracting solvent	Mobile phase consisting of methanol and water containing 0.1% orthophosphoric acid in the ratio of 50:50 v/v at a flow rate of 0.6 mL/min. Isocratic mode	YMC ODS C18 column	UV detection 267 nm	Concentration range of 20.41–2041.14 ng/mL with correlation coefficient 0.9994. The lower limit of quantification for CRZ in plasma was 20 ng/mL. No endogenous substances were found to interfere with the peaks of drug and internal standard The intra– and inter–day precision was <9.0% and the accuracy ranged from 97% to 112% over the linear range	[30]
21	Reverse–phase high-performance liquid chromatography	Crizotinib	Bulk form	Methanol and Sodium Phosphate Buffer 10 mmph 6.5 mixed in a proportion of 85:15 v/v	Prime's C18 column (Length: 250 nm, Diameter: 4.6 nm, Particle size:5 μ)	—	Concentration range of 10–50 μg/mL with the correlation coefficient of 0.998	[31]
22	NP–HPLC	Crizotinib	—	Mobile phase comprising a mixture of n–hexane–iso–Pr alc.–methanol–diethyl amine (40:30:30:0.5 v/v/v/v), pumped at a flow rate of 1.0 mL/min. S & R enantiomers, eluted at the retention times of 4.9 and 6.1 min, respectively isocratic	Chiralcel OD–H (25 × 0.46 cm, 5 μ) column	UV-Visible detector, 268 nm	10–200 μg/mL R2 0.999(S), 0998(R) For S & R enantiomers intra–day precision were found to be 0.41 and 0.61, respectively. inter–day precision was found to be %pooled RSD were found to be 0.96 and 0.61 for S & R enantiomers, respectively	[32]

5. ADME profiles of drug substances and excipients [6]

5.1 Uses, applications

Crizotinib (PF-02341066) is an anticancer drug approved by the US-FDA for the treatment of patients with metastatic non-small cell lung cancer (NSCLC) whose tumors are anaplastic lymphoma kinase (ALK) or ROS1-positive.

5.2 Absorption

The peak plasma concentrations of crizotinib reached between 4h post single oral dose. The steady-state concentrations reached within 15 days. The crizotinib can be given with or without food. Mean bioavailability is about 43% [6,33,34]. In pediatric patients with ALCL, crizotinib steady-state exposure increased proportionally with dose. Mean crizotinib AUC decreased by 35% and C_{max} decreased by 27% in patients with severe hepatic impairment. In patients with severe renal impairment, the AUC of crizotinib increased by 79% and the C_{max} increased by 34%.

5.3 Distribution

The Steady sate volume of distribution of crizotinib is 1772L. Protein binding of crizotinib is about 90%. In vitro data suggest that crizotinib is a substrate for P-glycoprotein. The blood-to-plasma concentration ratio is approximately 1 [6].

5.4 Metabolism

Crizotinib metabolism is mainly via the cytochrome P450 isoenzymes CYP3A4 and CYP3A5. The main metabolic pathways are oxidation (to crizotinib lactam) and 0-dealkylation, with subsequent phase 2 conjugation of O-dealkylation metabolites.

5.5 Elimination

The mean apparent plasma terminal half-life of crizotinib is 42h following single dose of crizotinib in patients. The mean apparent clearance (CL/F) of crizotinib was lower at steady-state (60L/h) after 250mg twice daily than after a single 250mg oral dose (100L/h) Following administration of a single oral 250mg dose of radiolabeled crizotinib dose to healthy subjects, 63% of the administered dose was recovered in feces and 22% in urine.

5.6 Pharmacological effects

Crizotinib is a receptor tyrosine kinases inhibitor (RTKs) including; Hepatocyte Growth Factor Receptor (HGFR, c-Met), anaplastic lymphoma kinase (ALK), ROS1 (c-ros), and Recepteur d'Origine Nantais (RON) [6,35,36]. Receptor tyrosine kinases play important roles in cell proliferation, migration, metabolism, differentiation, and survival. The enhanced RTK activities have been implicated in the development and progression of several types of cancer. Crizotinib is approved for the treatment of the subset of non-small-cell lung cancers (NSCLC) with rearrangements involving ALK. Translocations can affect the ALK gene resulting in the expression of onco-genic fusion proteins. The formation of ALK fusion proteins results in activation and dysregulation of the gene's expression and signaling which can contribute to increased cell proliferation and survival in tumors expressing these proteins. Crizotinib demonstrated concentration-dependent inhibition of ALK, ROS1, and c-Met phosphorylation in cell-based assays using tumor cell lines and demonstrated antitumor activity in mice bearing tumor xenografts that expressed echinoderm microtubule-associated protein-like 4 (EML4)- or nucleophosmin (NPM)-ALK fusion proteins or c-Met [6,37]. Christensen et al., reported that crizotinib potently inhibited cell proliferation and induced apoptosis in in ALK-positive ALCL-derived cell lines. These data further confirmed by in vivo data obtained in an ALCL-derived mouse model showed complete regression of the tumor at a dose of 100 mg/kg once daily [36]. The reported adverse reactions associated with crizotinib were; Hepatotoxicity, Interstitial Lung Disease/Pneumonitis, QT Interval Prolongation, Bradycardia, Vision disorders, Gastrointestinal toxicities.

5.7 Drug-drug interactions

In healthy subjects, the co-administration of crizotinib (single, oral dose) with the ketoconazole, strong CYP3A inhibitor, or resulted increased crizotinib plasma AUC by 3.2-fold [38]. Coadministration of crizotinib with Rifampin, strong CYP3A inducer, also decreased crizotinib steady-state AUC and C_{max} by 84% and 79%, respectively. However, the coadministration of dexamethasone with crizotinib has no effect on crizotinib exposure or efficacy [39]. The coadministration of crizotinib with ketoconazole, a strong CYP3A inhibitor, increased crizotinib AUC by 216% and C_{max} by 44%. While coadministration of crizotinib (250 mg orally twice daily for 28 days) with benzodiazepine midazolam, CYP3A substrate, increased the AUC of midazolam by 3.7-fold.

The aqueous solubility of crizotinib is dependent on pH, so higher pH resulted in decreased solubility and consequently drugs that increased pH may reduce the crizotinib solubility. Population PK analysis indicated that concomitant proton pump inhibitor administration decreased the absorption rate of crizotinib [40]. Shu et al., reported that entecavir has the potential to competitively inhibit crizotinib transport by organic cation transporter 2 (OCT2) in kidney, resulting in increased crizotinib exposure and reduced elimination in patients with NSCLC as well as increasing the incidences of crizotinib adverse reactions [41].

Zhou et al., reported that crizotinib can competitively inhibiting the transport function of ABCB1. This property could be utilized in enhancing the efficacy of chemotherapeutic drugs in ABCB1-overexpressing multidrug resistance (MDR) cells. These findings support potential usefulness of combining crizotinib with other conventional chemotherapeutic drugs in combating MDR in cancer chemotherapy [42].

6. Methods of chemical synthesis

6.1 Preparative chemical methods

The U.S. Patent No. 7858643 described method [see Scheme 1] involves deprotection of compound of Formula A in presence of HCl/dioxane to obtain compound of Formula B followed by Suzuki coupling with pyrazole-boc-piperidine compound of Formula IIIa to obtain compound of Formula IV, then deprotection of the Formula IV with 4 M hydrochloride and finally purified on a reverse phase C-18 preparative HPLC eluting with acetonitrile/water/0.1% acetic acid and lyophilized to obtain crizotinib as a acetate salt. The '643 patent discloses final obtained compound having 96.4% optical purity [43].

Cui et al. described a method for total synthesis of crizotinib outlined in Scheme 2. In this method, Boronic ester 6 was prepared via alkylation of 3 with piperidinyl derivative 2 followed by palladium coupling with 5. The compound 7 was prepared with the same synthetic route as the racemic analogue 5-bromo-3-(1-(2,6-dichloro-3-fluorophenyl)ethoxy) pyridin2-amine (Compound 22 as shown in Scheme 3) in published article). (S)-1-(2,6-dichloro-3-fluorophenyl) ethanol was obtained via a biotransformation method. Mitsunobu reaction of (S)-1-(2,6-dichloro-3-fluorophenyl) ethanol with compound 18 (compound in the article) provided (R)-3-(1-(2,-6-dichloro-3-fluorophenyl)ethoxy)-2-nitropyridine with >99.5% which was

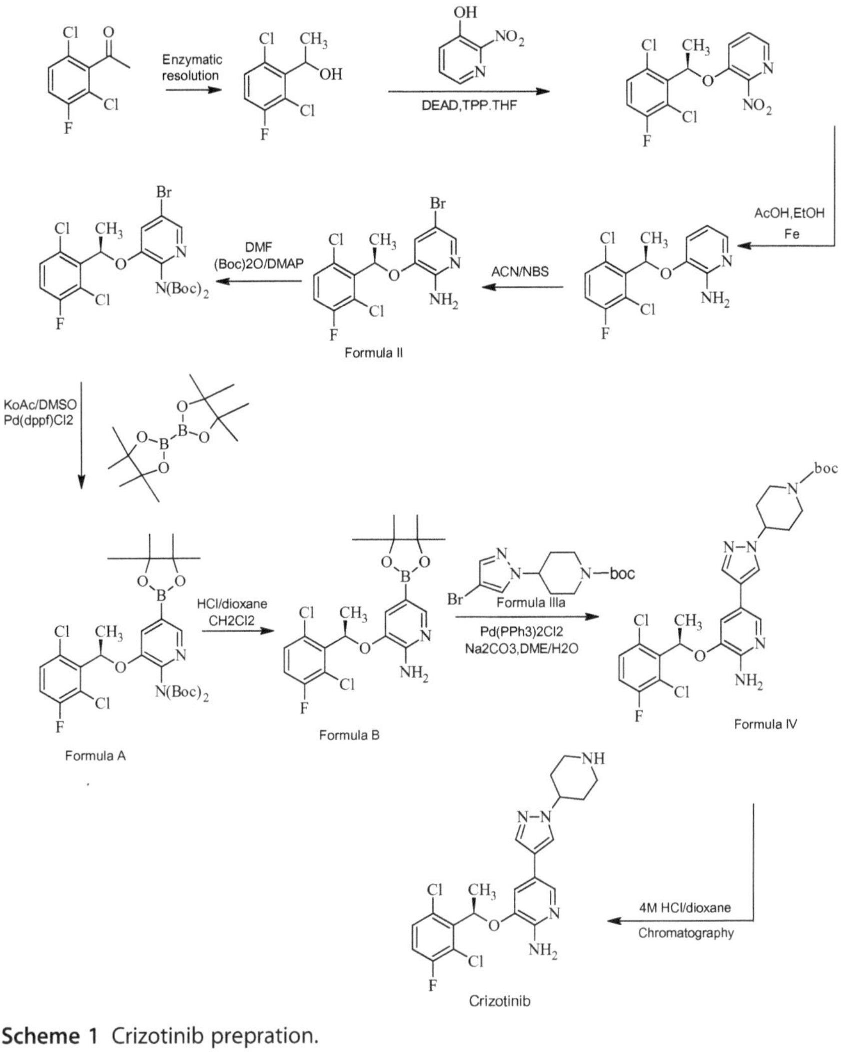

Scheme 1 Crizotinib prepration.

detailed in the published article. Suzuki coupling reaction of 7 with 6 followed by deprotection with HCl generated crizotinib in good yield [44].

The Chinese Patent Application No's: 101735198 and 102584795 reported a process for preparation of crizotinib [Scheme 3], which involves resolution of S-isomer of 1-(2,6-dichloro-3-fluorophenyl) ethanol and coupling of bromo pyridine intermediate of Formula II and pyrazole boronic ester compound of Formula III in presence of palladium catalyst and base to yield compound of Formula and followed by deprotection to yield crizotinib [43].

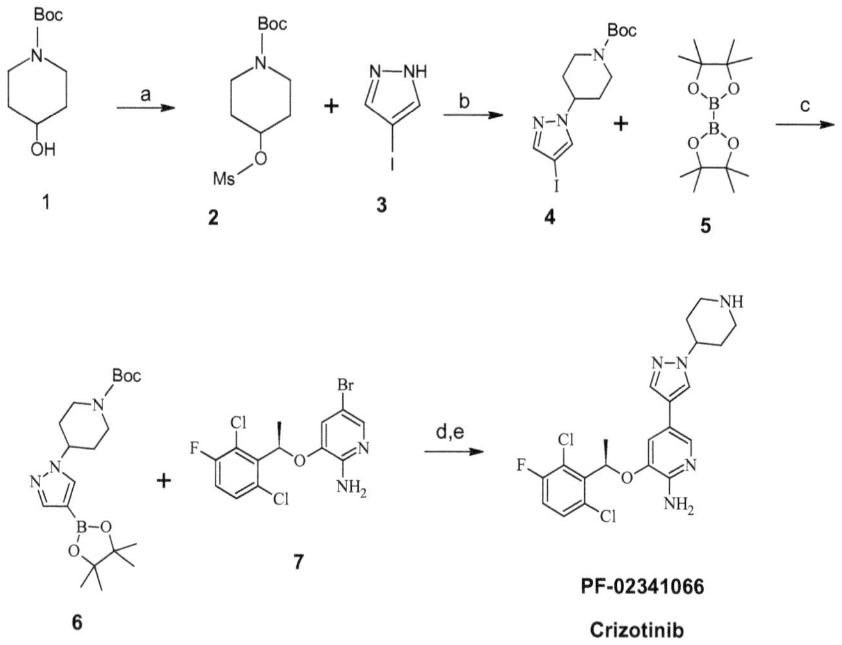

Scheme 2 Reagents and conditions: (A) MsCl, Et$_3$N, CH$_2$Cl$_2$; (B) NaH, DMF, 100 °C, overnight; (C) Pd(Ph$_3$P)$_2$Cl$_2$, KOAc, DMSO, 80 °C, 2 h; (D) Pd(dppf)Cl$_2$, Cs$_2$CO$_3$, DME/H$_2$O, 90 °C, 3 h; (E) 4 N HCl in dioxane, CH$_2$Cl$_2$, 0 °C, 4 h.

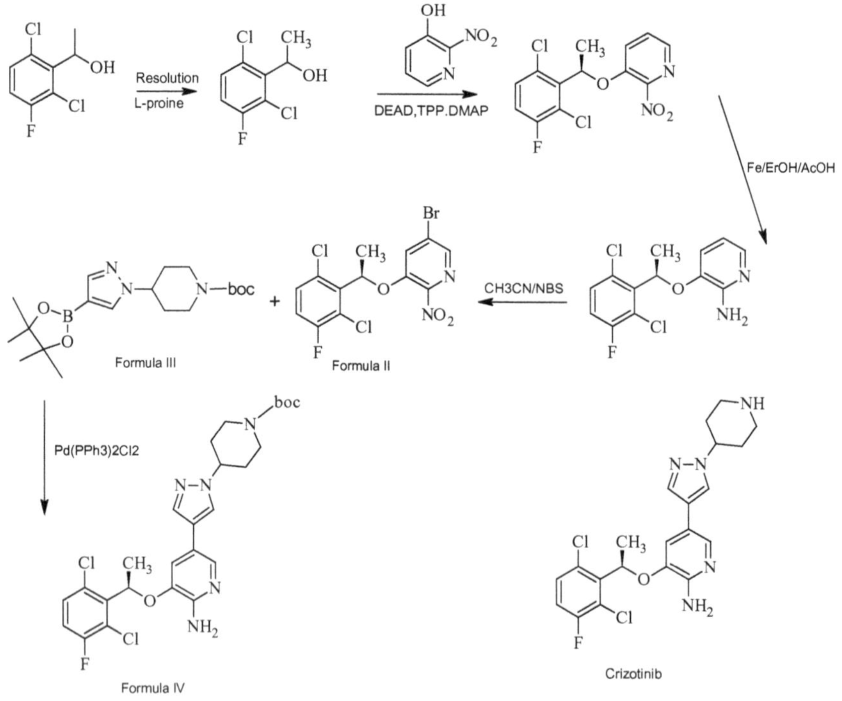

Scheme 3 Crizotinib prepration.

The Chinese Patent application No: 103420987 described a method for preparation of crizotinib, which involves Suzuki coupling of amine protected compound of Formula C [see Fig. 1] with pyrazole-boc-piperidine compound of Formula IIIa [see Scheme 1] in presence of palladium catalyst and base to yield amine protected compound of Formula IV [see Scheme 1] and followed by deprotection to yield crizotinib [43].

Patent publication No. WO 2014124594 discloses a process for preparation of crizotinib, which involves Suzuki coupling of amine protected compound of Formula D [see Fig. 1] with pyrazole boronic ester compound of Formula III [see Scheme 3] in presence of palladium catalyst and base to yield amine protected compound of Formula IV and followed by deprotection to yield crizotinib [43].

U.S. Patent No. 9604966B2 Li et al. described a modified method for crizotinib preparation involve preparing the compound of formula a comprises Suzuki coupling reaction of the compound of as formula b and the compound of formula c to produce the compound of formula a [see Fig. 1 [45].

Simone et al. [46] also reported method for preparing Crizotinib hydrochloride salt in crystalline. Here is an example method followed; Crizotinib free-base (205 mg, 0.4 mmol) was suspended in 4 mL ethylacetate and heated to 50 °C. Hydrochloric acid (420 µ; 1.25 M-solution in ethanol, 0.4 mmol, 1 equiv.) was added dropwise at 50 °C. The resulting suspension was cooled to 25 °C and afterward stored at 25 °C for 2 days. The resulting precipitate was isolated by filtration and washed with ethylacetate (5 mL). The product was dried for 5 days under normal pressure at 25 °C to yield 183 mg of Crizotinib hydrochloride form HI (84.6%, 0.4 mmol) as a white powder. The authors also extended the invention to include other crizotinib salts such as Crizotinib phosphate, Crizotinib sulphate, Crizotinib acetate, Crizotinib maleate, Crizotinib fumarate, Crizotinib L-tartrate, Crizotinib citrate, Crizotinib p-toluene sulfonate, Crizotinib succinate, Crizotinib formate, Crizotinib besilate, Crizotinib ethane- 1,2-disulfonate and Crizotinib L-malate.

Chava et al. reported method for industrial preparation of crizotinib or acid addition salt thereof of Formula I, comprising; (A) reacting a compound of Formula II with a compound of Formula III in presence of a base (bl) and palladium catalyst in a suitable solvent (SI) to obtain a compound of Formula IV wherein and R2 are independently selected from the group consisting of hydroxy, optionally substituted -C alkoxy and optionally substituted -Cs alkyl, or Ri and R2 together form an optionally substituted -C3 alkylenedioxy group or an optionally substituted C6 aryldioxy group;

Pi, P2 and P3 independently represent hydrogen or a suitable amino protecting group; and L represents halogen; (B) deprotecting the obtained compound of Formula IV in presence of a suitable acid to obtain corresponding acid addition salt of crizotinib of Formula V; (C) neutralizing the obtained acid addition salt of Formula V in presence of a suitable base (b2) to obtain crizotinib of Formula I [47].

Formula I   Formula II   Formula III   Formula IV

Xu et al., developed large scale production of crizotinib through catalytic hydrogenation of aryl nitro group in the (R)-3-(1-(2,6-dichloro-3-fluorophenyl)ethoxy)-2-nitropyridine leading to the desired product (R)-3-[1-(2,6-dichloro-3-fluorophenyl)ethoxy]pyridin-2-amine from which the crizotinib was finally synthesized [48].

References

[1] D.R. Tonkin, K. Haskins, Regulatory T cells enter the pancreas during suppression of type 1 diabetes and inhibit effector T cells and macrophages in a TGF-beta-dependent manner, Eur. J. Immunol. 39 (5) (2009) 1313–1322.
[2] A. Sahu, K. Prabhash, V. Noronha, A. Joshi, S. Desai, Crizotinib: a comprehensive review, South Asian J Cancer 2 (2) (2013) 91–97.
[3] N.C. Schlegel, O.M. Eichhoff, S. Hemmi, S. Werner, R. Dummer, K.S. Hoek, Id2 suppression of p15 counters TGF-beta-mediated growth inhibition of melanoma cells, Pigment Cell Melanoma Res. 22 (4) (2009) 445–453.
[4] R.W. Sparidans, S.C. Tang, L.N. Nguyen, A.H. Schinkel, J.H. Schellens, J.H. Beijnen, Liquid chromatography-tandem mass spectrometric assay for the ALK inhibitor crizotinib in mouse plasma, J. Chromatogr. B Analyt Technol. Biomed. Life Sci. 905 (2012) 150–154.
[5] Chemical Book, Available from: https://www.chemicalbook.com/ProductChemical PropertiesCB12473904_EN.htm.
[6] US Food and Drug Administration, XALKORI® (Crizotinib) Package Insert, 2021. https://www.accessdata.fda.gov/drugsatfda_docs/label/2021/202570s030lbl.pdf.

[7] I.A. Darwish, J.M. Alshehri, N.Z. Alzoman, N.Y. Khalil, H.M. Abdel-Rahman, Charge-transfer reaction of 1,4-benzoquinone with crizotinib: spectrophotometric study, computational molecular modeling and use in development of microwell assay for crizotinib, Spectrochim. Acta A Mol. Biomol. Spectrosc. 131 (2014) 347–354.

[8] I.A. Darwish, J.M. Alshehri, N.Z. Alzoman, N.Y. Khalil, H.M. Abdel-Rahman, Charge-transfer reaction of chloranilic acid with crizotinib: spectrophotometric study, computational modeling and use in development of microwell assay for crizotinib, J. Sol. Chem. 43 (2014) 1282–1295.

[9] N.Z. Alzoman, J.M. Alshehri, I.A. Darwish, N.Y. Khalil, H.M. Abdel-Rahman, Charge-transfer reaction of 2,3-dichloro-1,4-naphthoquinone with crizotinib: spectro-photometric study, computational molecular modeling and use in development of microwell assay for crizotinib, Saudi Pharm. J. 23 (1) (2015) 75–84.

[10] D. Tawa, A novel 96-microwell-based high-throughput spectrophotometric assay for pharmaceutical quality control of crizotinib, a novel potent drug for the treatment of non-small cell lung cancer, Braz. J. Pharm. Sci. 51 (2) (2015).

[11] W. Hany, A.H.B. Darwish, I.A. Darwish, Enhanced spectrofluorimetric determination of the multitargeted tyrosine kinase inhibitor, crizotinib, in human plasma via micelle-mediated approach, Trop. J. Pharm. Res. 15 (10) (2016) 2209–2217.

[12] A.S. Abdelhameed, A.M. Alanazi, A.H. Bakheit, H.W. Darwish, H.A. Ghabbour, I.A. Darwish, Fluorescence spectroscopic and molecular docking studies of the binding interaction between the new anaplastic lymphoma kinase inhibitor crizotinib and bovine serum albumin, Spectrochim. Acta A Mol. Biomol. Spectrosc. 171 (2017) 174–182.

[13] M.M. Al-Shehri, A.S. El-Azab, M.A. El-Gendy, M.A. Hamidaddin, I.A. Darwish, Synthesis of hapten, generation of specific polyclonal antibody and development of ELISA with high sensitivity for therapeutic monitoring of crizotinib, PLoS One. 14 (2) (2019) e0212048.

[14] M.S. Roberts, D.C. Turner, A. Broniscer, C.F. Stewart, Determination of crizotinib in human and mouse plasma by liquid chromatography electrospray ionization–tandem mass spectrometry (LC-ESI-MS/MS), J. Chromatogr. B Analyt Technol. Biomed Life Sci. 960 (2014) 151–157.

[15] F. Qiu, Y. Gu, T. Wang, Y. Gao, X. Li, X. Gao, et al., Quantification and pharmaco-kinetics of crizotinib in rats by liquid chromatography-tandem mass spectrometry, Biomed. Chromatogr. 30 (6) (2016) 962–968.

[16] F. Zhao, Y. Wei, Y. Yan, H. Liu, S. Zhou, B. Ren, et al., Determination of crizotinib in mouse tissues by LC-MS/MS and its application to a tissue distribution study, Int. J. Anal. Chem. 2020 (2020) 8837254.

[17] M.W. Ni, J. Zhou, H. Li, W. Chen, H.Z. Mou, Z.G. Zheng, Simultaneous determi-nation of six tyrosine kinase inhibitors in human plasma using HPLC-Q-Orbitrap mass spectrometry, Bioanalysis 9 (12) (2017) 925–935.

[18] M. Herbrink, N. de Vries, H. Rosing, A.D.R. Huitema, B. Nuijen, J.H.M. Schellens, et al., Development and validation of a liquid chromatography-tandem mass spectrom-etry analytical method for the therapeutic drug monitoring of eight novel anticancer drugs, Biomed. Chromatogr. 32 (4) (2018).

[19] C. Merienne, M. Rousset, D. Ducint, N. Castaing, K. Titier, M. Molimard, et al., High throughput routine determination of 17 tyrosine kinase inhibitors by LC-MS/MS, J. Pharm. Biomed. Anal. 150 (2018) 112–120.

[20] R. Reis, L. Labat, M. Allard, P. Boudou-Rouquette, J. Chapron, A. Bellesoeur, et al., Liquid chromatography-tandem mass spectrometric assay for therapeutic drug monitor-ing of the EGFR inhibitors afatinib, erlotinib and osimertinib, the ALK inhibitor crizotinib and the VEGFR inhibitor nintedanib in human plasma from non-small cell lung cancer patients, J. Pharm. Biomed. Anal. 158 (2018) 174–183.

[21] X. Qi, L. Zhao, Q. Zhao, Q. Xu, Simple and sensitive LC-MS/MS method for simultaneous determination of crizotinib and its major oxidative metabolite in human plasma: application to a clinical pharmacokinetic study, J. Pharm. Biomed. Anal. 155 (2018) 210–215.

[22] G.D.M. Veerman, M.H. Lam, R.H.J. Mathijssen, S.L.W. Koolen, P. de Bruijn, Quantification of afatinib, alectinib, crizotinib and osimertinib in human plasma by liquid chromatography/triple-quadrupole mass spectrometry; focusing on the stability of osimertinib, J. Chromatogr. B Analyt Technol. Biomed. Life Sci. 1113 (2019) 37–44.

[23] A. van Veelen, R. van Geel, R. Schoufs, Y. de Beer, L.M. Stolk, L.E.L. Hendriks, et al., Development and validation of an HPLC-MS/MS method to simultaneously quantify alectinib, crizotinib, erlotinib, gefitinib and osimertinib in human plasma samples, using one assay run, Biomed. Chromatogr. (2021) e5224.

[24] H.M. Maher, A. Almomen, N.Z. Alzoman, S.M. Shehata, A.A. Alanazi, Development and validation of UPLC-MS/MS method for the simultaneous quantification of anaplastic lymphoma kinase inhibitors, alectinib, ceritinib, and crizotinib in Wistar rat plasma with application to bromelain-induced pharmacokinetic interaction, J. Pharm. Biomed. Anal. 204 (2021) 114276.

[25] Y. Mukai, A. Wakamoto, T. Hatsuyama, T. Yoshida, H. Sato, A. Fujita, et al., An LC-MS/MS method for the simultaneous determination of afatinib, alectinib, ceritinib, crizotinib, dacomitinib, erlotinib, gefitinib, and osimertinib in human serum, Ther. Drug Monit. (2021).

[26] V. Bontha, E. Suman, G. Sreekanth, Estimation of crizotinib in capsule dosage form by RP-HPLC, IOSR J. Pharm. Biol. Sci. 11 (6) (2016) 93–103.

[27] X. Huang, J. Cai, X. Wang, LC-MS determination of crizotinib in rat plasma and its application to a pharmacokinetic study, Lat. Am. J. Pharm. 33 (7) (2014) 1188–1192.

[28] N.Y. Khalil, T.A. Wani, I.A. Darwish, A.-R.A. Al-Majed, Highly sensitive HPLC method with non-extractive sample preparation and fluorescence detection for determination of crizotinib in human plasma, Lat. Am. J. Pharm. 33 (6) (2014) 1019–1026.

[29] J. Bandla, S. Ganapaty, Stability indicating UPLC method development and validation for the determination of crizotinib in pharmaceutical dosage forms, Int. J. Pharm. Sci. Res. 9 (4) (2018) 1493–1498.

[30] B.N. Revu, S.R. Atla, G.S. Dannana, Quantitative determination of crizotinib in human plasma with highperformance liquid chromatography and ultraviolet detection, Asian J. Pharm. Clin. Res. 12 (2) (2019) 363–367.

[31] P.B. Jadhav, V. Shejwal, Development and validation of an RP-HPLC method for crizotinib, Int. J. Pharm. Pharm. Res. 9 (2) (2017) 100–106.

[32] B. Vijayakumar, B.L. Samhitha, A. Ramu, M.V. Krishna, G. Sreekanth, Enantioselective analysis of crizotinib by chiral LC method, Pharma Chem. 9 (22) (2017) 18–24.

[33] H. Xu, M. O'Gorman, T. Boutros, N. Brega, C. Kantaridis, W. Tan, et al., Evaluation of crizotinib absolute bioavailability, the bioequivalence of three oral formulations, and the effect of food on crizotinib pharmacokinetics in healthy subjects, J. Clin. Pharmacol. 55 (1) (2015) 104–113.

[34] S.H. Ou, Crizotinib: a novel and first-in-class multitargeted tyrosine kinase inhibitor for the treatment of anaplastic lymphoma kinase rearranged non-small cell lung cancer and beyond, Drug Des. Devel. Ther. 5 (2011) 471–485.

[35] H.Y. Zou, Q. Li, J.H. Lee, M.E. Arango, S.R. McDonnell, S. Yamazaki, et al., An orally available small-molecule inhibitor of c-Met, PF-2341066, exhibits cytoreductive antitumor efficacy through antiproliferative and antiangiogenic mechanisms, Cancer Res. 67 (9) (2007) 4408–4417.

[36] J.G. Christensen, H.Y. Zou, M.E. Arango, Q. Li, J.H. Lee, S.R. McDonnell, et al., Cytoreductive antitumor activity of PF-2341066, a novel inhibitor of anaplastic lymphoma kinase and c-Met, in experimental models of anaplastic large-cell lymphoma, Mol. Cancer Ther. 6 (12 Pt. 1) (2007) 3314–3322.

[37] S. Yamazaki, P. Vicini, Z. Shen, H.Y. Zou, J. Lee, Q. Li, et al., Pharmacokinetic/pharmacodynamic modeling of crizotinib for anaplastic lymphoma kinase inhibition and antitumor efficacy in human tumor xenograft mouse models, J. Pharmacol. Exp. Ther. 340 (3) (2012) 549–557.

[38] H. Xu, M. O'Gorman, W. Tan, N. Brega, A. Bello, The effects of ketoconazole and rifampin on the single-dose pharmacokinetics of crizotinib in healthy subjects, Eur. J. Clin. Pharmacol. 71 (12) (2015) 1441–1449.

[39] S. Lin, D.J. Nickens, M. Patel, K.D. Wilner, W. Tan, Clinical implications of an analysis of pharmacokinetics of crizotinib coadministered with dexamethasone in patients with non-small cell lung cancer, Cancer Chemother. Pharmacol. 84 (1) (2019) 203–211.

[40] G. Hamilton, B. Rath, O. Burghuber, Pharmacokinetics of crizotinib in NSCLC patients, Expert Opin. Drug Metab. Toxicol. 11 (5) (2015) 835–842.

[41] W. Shu, L. Ma, X. Hu, M. Zhang, W. Chen, W. Ma, et al., Drug-drug interaction between crizotinib and entecavir via renal secretory transporter OCT2, Eur. J. Pharm. Sci. 142 (2020) 105153.

[42] W.J. Zhou, X. Zhang, C. Cheng, F. Wang, X.K. Wang, Y.J. Liang, et al., Crizotinib (PF-02341066) reverses multidrug resistance in cancer cells by inhibiting the function of P-glycoprotein, Br. J. Pharmacol. 166 (5) (2012) 1669–1683.

[43] G. Chava, Indukuri, Moturu, Karuturi, inventor; Laurus Labs Private Limited, assignee, A Process for the Preparation of Crizotinib or an Acid Addition Salt Thereof 2015, 2015.

[44] J.J. Cui, M. Tran-Dubé, H. Shen, M. Nambu, P.P. Kung, M. Pairish, et al., Structure based drug design of crizotinib (PF-02341066), a potent and selective dual inhibitor of mesenchymal-epithelial transition factor (c-MET) kinase and anaplastic lymphoma kinase (ALK), J. Med. Chem. 54 (18) (2011) 6342–6363.

[45] Y.L. Taizhou, J.Q. Taizhou, D.C. Taizhou, Crizotinib Preparation Method, 2017. US 9604966B2.

[46] E. Simone, B. Wolfgang, Guserle, Richard, L. Frank, Crizotinib hydrochloride salt in crystalline, 2013 international publication no. wo 2013/181251 Al, 2013.

[47] S. Chava, S.R.A. Gorantla, V.S.K. Indukuri, V.R.K.M. Moturu, V.V.R. Karuturi, A Process for the Preparation of Crizotinib or an Acid Addition Salt Thereof, 2015. International publication number WO2015/107553A2.

[48] F. Xu, J. Chen, X. Xie, P. Cheng, Z. Yu, W. Su, Synthesis of a crizotinib intermediate via highly efficient catalytic hydrogenation in continuous flow, Org. Proc. Res. Dev. 24 (10) (2020) 2252–2259.

CHAPTER THREE

Remdesivir

Ahmed H. Bakheit[a,b], Hany Darwish[a], Ibrahim A. Darwish[a], and Ahmed I. Al-Ghusn[a]

[a]Department of Pharmaceutical Chemistry, College of Pharmacy, King Saud University, Riyadh, Kingdom of Saudi Arabia
[b]Department of Chemistry, Faculty of Science and Technology, Al-Neelain University, Khartoum, Sudan

Contents

Profiles of Drug Substances, Excipients, and Related Methodology, Volume 48
ISSN 1871-5125
https://doi.org/10.1016/bs.podrm.2022.11.003

1. Description

Remdesivir (GS-5734) is the first medication for severe coronavirus disease that was approved by the FDA in 2019 (COVID-19). It is a novel nucleoside analog with broad antiviral activity against a wide range of RNA viruses, including ebolavirus (EBOV) and respiratory pathogens like Middle East respiratory syndrome coronavirus (MERS-CoV), SARS-CoV, and SARS-CoV-2 [1]. Remdesivir, a single diastereomer monophosphoramidate as a prodrug, is converted into the active form GS-441524 after administration. GS-441524 competes with ATP for RNA incorporation and inhibits viral RNA-dependent RNA polymerase. This stops RNA transcription and reduces viral RNA replication [2].

1.1 Nomenclature

1.1.1 IUPAC name

- 2-ethylbutyl (2S)-2-[[[(2R,3S,4R,5R)-5-(4-aminopyrrolo[2,1-f][1,2,4] triazin-7-yl)-5-cyano-3,4-dihydroxyoxolan-2-yl]methoxy-phenoxyphosphoryl]amino]propanoate [3].
- 2-ethylbutyl (2S)-2-{[(S)-[(2R,3S,4R,5R)-5-{4-aminopyrrolo[2,1-f] [1,2,4]triazin-7-yl}-5-cyano-3,4-dihydroxyoxolan-2-yl]methoxy (phenoxy)phosphoryl]amino}propanoate [4].

1.1.2 Non-proprietary names (generic names) [5]

- Remdesivir
- Remdesivirum
- GS 5734
- UNII-3QKI37EEHE
- 2-ethylbutyl (2S)-2-(((2R, 3S, 4R, 5R)-5-(4-aminopyrrolo(2,1-f) (1,2,4)triazin-7-yl)-5-cyano-3,4-dihydroxytetrahydrofuran-2-yl) methoxy) (phenoxy) phosphoryl) amino) propanoate [6].

- L–Alanine, N-((S)-hydroxyphenoxyphosphinyl)-, 2-ethylbutyl ester, 6–ester with 2-C-(4-aminopyrrolo(2,1-f)(1,2,4)triazin-7-yl)-2,5-anhydro-d-altrononitrile [7].
- (2S)-2-{(2R,3S,4R,5R)-[5-(4-aminopyrrolo[2,1-f][1,2,4]triazin-7-yl)-5-cyano-3,4-dihydroxy-tetrahydro-furan-2-ylmethoxy]phenoxy-(S)-phosphorylamino}propionic acid 2-ethyl-butyl ester [7].
- 2-ethylbutyl N-[(S)-{[(2R,3S,4R,5R)-5-(4-aminopyrrolo[2,1-f][1,2,4] triazin-7-yl)-5-cyano-3,4-dihydroxytetrahydrofuran-2-yl]methoxy} (phenoxy)phosphoryl]-L-alaninate [7].
- 2-Ethylbutyl (2S)-2-{[(S)-{[(2R,3S,4R,5R)-5-(4-aminopyrrolo[2,1-f] [1,2,4]triazin-7-yl)-5-cyano-3,4-dihydroxytetrahydro-2-furanyl] methoxy}(phenoxy)phosphoryl] amino}propanoate [8].

1.1.3 Proprietary names (brand names)
- Veklury [9].

1.2 Formula
1.2.1 Empirical formula, molecular weight, CAS number
The empirical formula, molecular weight, and CAS number of vandetanib are shown in Table 1.

1.2.2 Structural formula
A representation of the two-dimensional structural formula of Remdesivir may be seen in Fig. 1.

1.2.3 Elemental analysis
See Table 2.

Table 1 Empirical formula, molecular weight, CAS number [5].

Compounds	Empirical formula	Molecular weight	CAS number
Remdesivir	$C_{27}H_{35}N_6O_8P$	602.6	1809249–37–3

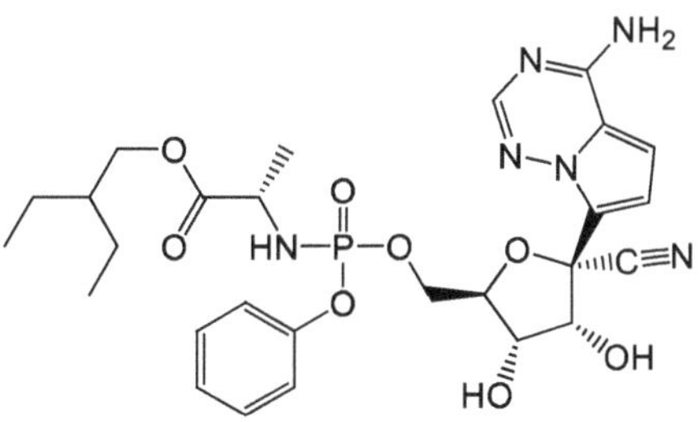

Fig. 1 Chemical structure of Remdesivir.

Table 2 Remdesivir's theoretical
elemental analysis and composition
were presented in Table 2 [3,5].

Elements	Remdesivir (%)
C	53.82
H	5.85
N	13.95
O	21.24
P	5.14

2. Methods of synthesis

Siegel et al. [10] synthesized a single Sp phosphoramidate prodrug (Remdesivir) by combining comp **6** with glycosylation via metal-halogen exchange of the bromo-base **3** followed by addition into the ribolactone **2** (Scheme 1). Two conditions were identified that resulted in the creation of the required C-C bond. The first condition (a) was achieved by adding excess n–BuLi to a solution of TMSCl and **3**, which was expected to result in lithium-halogen exchange following silyl protection of the acidic 6N protons. After adding this in situ produced reagent to the ribolactone **2**, **4** was obtained in a 25% yield. In the alternate condition (b), sodium hydride and 1,2-bis(chlorodimethylsilyl)ethane were used to guard the 6N atoms, followed by lithium-halogen exchange and addition to the lactone to get **4** in 60% yield. Both conditions were inefficient since the yields were highly variable and greatly reliant on the cryogenic temperatures and rate of n–BuLi addition required for the transformation. Additionally, we observed premature quenching and reduction of lithio base, which was explained as a result of alpha deprotonation to the lactone under very basic circumstances. Compound **4** was separated as a mixture of 1'-isomers that were then used in the subsequent 1'-cyanation procedure to obtain the primary product, -anomer **5**. After removing the three benzyl protecting groups to obtain **6**, the diastereomeric combination of the phosphoramidoyl chloridate prodrug moiety **7** was linked to generate 1a in a yield of 21% as a 1:1 diastereomeric mixture.

The synthetic approach proposed by Gilead scientists for commercial manufacturing of Remdesivir entails protecting group modifications of

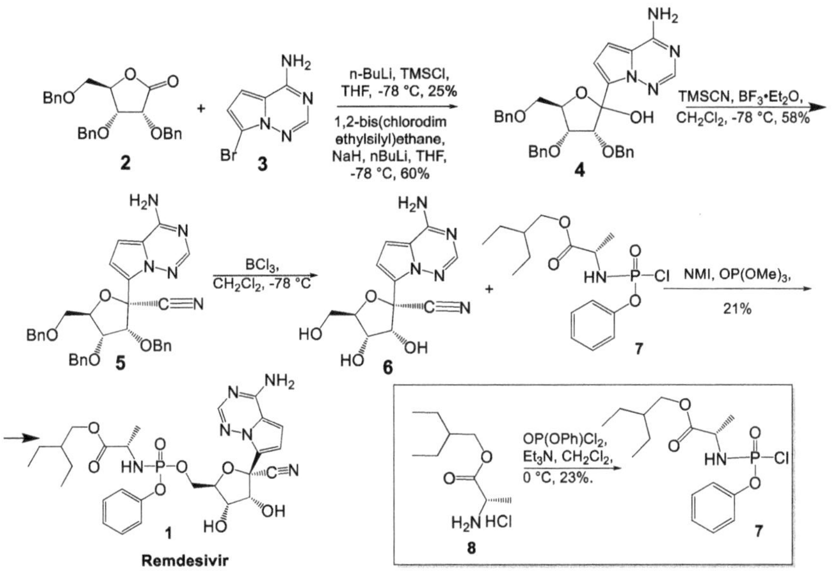

Scheme 1 Siegel route for synthesis of Remdesivir.

the sugar portion during the synthesis, allowing for further optimization [11,12]. The conversion of D-ribose (**2**) to 2,3,5-tri-O-benzyl-D-ribonolactone (**3**), followed by the addition of the modified nucleobase 4-amino-7-iodopyrrolo[2,1-f][1,2,4]triazine (**4**) to the lactone, yielded the C-glycosylated product **5**. BCl₃ was used to remove all three benzyl groups from the triol, and the two secondary alcohols were then protected as isopropylidene ketal **7** after the cyanation process to product **6**. When phosphate ester **8** is added to ketal **7** after hydrolysis, it forms Remdesivir (**1**) (Scheme 2).

Kumar Palli et al. [12] developed a short synthetic route to Remdesivir in seven longest linear steps with a total yield of 25% starting from D-ribonolactone **2** (Scheme 3A and B). The process involves silylating the primary hydroxyl group of commercially available D-ribonolactone (**2**, which can alternatively be produced in one step from D-ribose), using *tert*-butyldiphenyl silyl (TBDPS) to yield the corresponding silyl ether **3** in 84% yield. Additional protection of the remaining two secondary alcohols with allyl *tert*-butyl carbonate **4** in the presence of 2 mol% Pd(PPh₃)₄ resulted in 89% yield of diallylated ribonolactone **5** (Scheme 3A). Following that, lactone **5** was C-glycosylated with the protected nucleobase tert-butyl(7-bromopyrrolo [2,1-f] [1,2,4] triazin-4-yl) carbamate **6**, which produced an anomeric

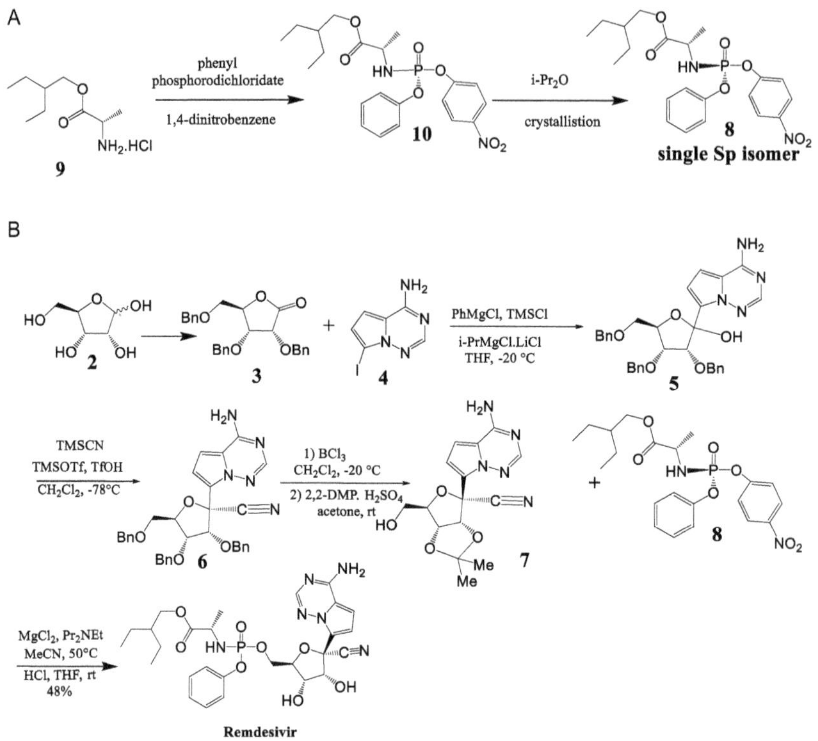

Scheme 2 (A) Synthesis of a single Sp isomer **8** from a *p*-nitrophenolate prodrug precursor **10** instead of phenyl phosphorodichloridate afforded. (B) Gilead route for the synthesis of Remdesivir (**1**, GS-5734).

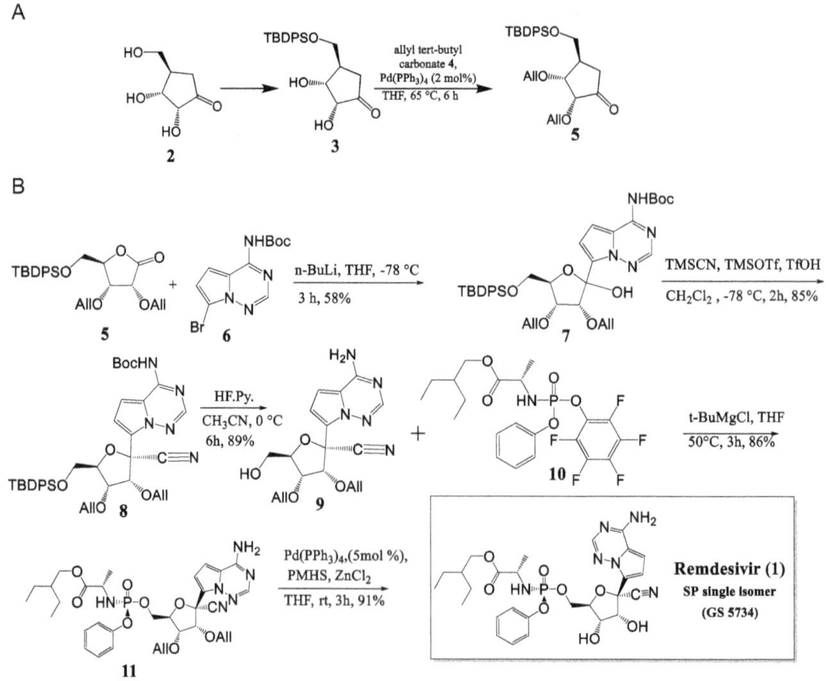

Scheme 3 (A) Synthesis of 5-O-TBDPS-2,3-O-diallyl ribonolactone **5**. (B) Total synthesis of remdesivir (**1**, *GS-5734*).

combination (2:1) of C-glycosylated product **7** in 58% yield in the presence of n–BuLi. Under standard conditions, cyanation of **7** produced cyano-glycoside **8** in an 85% yield with high selectivity for the target isomer (d.r. = 96:4, β:α). To insert the required phosphonate on the primary hydroxyl group, the silyl and Boc groups were deprotected in one-pot using HF. Pyridine to give amino-alcohol **9** in 89% yield. The following step was P-chiral phosphorylation, a critical step in the production of Remdesivir. The coupling reaction of **9** with the known chiral pentafluoro-phosphoramidate **10** in the presence of t–BuMgCl led to formation of the phosphoramidate ester **11** in an 86% yield as a single diastereoisomer.

3. Physical characteristics

3.1 Physical description

Remdesivir is made up of a non–hygroscopic crystalline solid that comes in a range of colors from white and off-white to yellow [13].

3.2 Dissociation constants

pK_a (strongest acidic/basic) = 10.23/0.65 [2].

3.3 Solubility characteristics

Remdesivir is soluble in ethanol and completely soluble in methanol. It is virtually insoluble in water and its aqueous solubility is pH dependent, increasing as the pH drops; It is very slightly soluble in water when pH is adjusted to pH 2 using HCl [13].

3.4 Partition coefficient

Log Po/w = 2.01 [2].

3.5 Particle morphology

3.5.1 Single crystal structure

Sekharan et al. [14] demonstrated a stable solid of Remdesivir form in a short period of time using the microcrystal electron diffraction (MicroED) technology and a cloud-based and artificial intelligence-based crystal structure prediction platform.

In agreement with experimental findings, the MicroED structures of Remdesivir forms II and IV were shown, and it was found that form

II is more stable at room temperature than form IV. Diffraction data were collected from ten distinct Remdesivir form II crystals, each of which covered ~30° of reciprocal space. To eliminate diffractions with a low signal-to-noise ratio, the resolution was limited to 0.900 Å. The combined data set contains 11 574 total diffractions and 3562 individual diffractions, with a 91% completeness rate and a Rint value of 0.2297. The observed symmetry of the diffraction intensity is 2/m Laue, indicating that the Remdesivir form II crystal is monoclinic. With the P21 space group, the unit-cell constants are a = 10.21(4) Å, b = 12.49(14) Å, c = 10.85(10) Å, α = 90°, β = 100.9(6)°, γ = 90°. Generally, the experimental structures were identified in the CSP landscape, which was used for validation by comparing the predicted structures from the calculation of XRDs. In the Remdesivir landscape, there are 35 crystal polymorphs predicted, with 22 belonging to the P21 space group, eight to the P212121 space group, and five to the P1 space group. Within a relative lattice energy gap of 10kJ mol^{-1}, only three crystal polymorphs (X1, X2, X3) belonging to the P21 space group are discovered. The comparison of predicted and experimental XRPD patterns, as well as the RMSD15 structural overlays, indicate that X1 and X2 are the experimental structures corresponding to forms IV (RMSD = 0.368) and II (RMSD = 0.441), respectively. In comparison to X1, the most dramatic stabilization occurs for X2, which loses over 5kJ mol^{-1} of energy to become more stable than X1 at room temperature, consistent with experimental data. At 300 K, the difference in the amount of energy between X1 and X2 is 0.76kJ mol^{-1}, which is within the CSP calculations' estimated range of 1.5kJ mol^{-1} for the energy difference. There are two polymorphs, and it's hard to tell which one is the most stable based on the results of the CSP test alone. The free energy calculations show that there isn't a stable form that hasn't been found yet. So, when choosing a stable solid form of Remdesivir, you should look at the competitive slurry tests between polymorphs II and IV. It looks like form II (X2) is the most stable solid form of Remdesivir.

Yu et al. [15] used A Bruker D8 Quest CCD diffractometer with Ga Kα (=1.341 Å) and a Bruker Apex-II CCD diffractometer with Mo Kα (= 0.7107 Å) to get the crystal data for RDV-I and RDV-II. On data reduction, the SAINT V8.38A program was employed. The absorption adjustment was carried out using the SADABS program's semi-empirical methods. Direct methods were used to figure out the crystal structures.

They were then refined with full-matrix least-squares methods with anisotropic thermal parameters for all non-hydrogen atoms on F^2 using SHELXL-2016. An isotropic riding model was used to refine hydrogen atoms in the location of calculation after they were placed in the position of calculation. Refinement of the crystal structure of RDV-I: The two alkane chains that were attached to C22A and C22B had a lot of conformational disorder. Their relative ratios of 0.45/0.55 and 0.47/0.53 were fine-tuned for the disordered areas. The structure is twinned around the (1 0 0) lattice direction. The twin matrix was suggested by the ROTAX program, and the HKLF 5 type file was made with the MAKE HKLF5 tool in the WinGX suite. Two twinned domains' ratios were fine-tuned to 0.58/0.42.

Single crystal X-ray structure analysis showed that RDV-I and RDV-II crystallize in the *P1* and *P2₁* groups, respectively (Table 3). RDV-asymmetric I's unit has two RDV molecules, whereas RDV-asymmetric II's unit contains only one RDV molecule. In a structure overlay, you can see that the two RDV molecules in RDV-I have the same conformation, but RDV-II has a very different conformation from that of RDV-I. They have a lot of flexibility because there are single bonds, which allows them to move around in different ways. This is especially true of the three bonds around the phosphorus centers that have single bonds.

3.5.2 X-ray powder diffraction pattern

Sahakijpijarn et al. [16] evaluated the powders' crystallinity using a benchtop X-ray diffraction apparatus (Rigaku Miniflex 600 II, Woodlands, TX, USA) equipped with primary monochromated radiation (Cu K radiation source, $\lambda = 1.54056$). The instrument was run at a 40 kV accelerating voltage and a current of 15 mA. Samples were placed into the sample holder and scanned in continuous mode at a scan speed of 2°/min and a dwell period of 2 s with a step size of 0.02 degree over a 2θ range of 5–40 degree. The X-ray diffraction patterns of Remdesivir powder formulations with thin film freezing (TFF) are provided different results. There were no strong peaks identified in any of the formulations, indicating that Remdesivir was amorphous following the (TFF) procedure. The drug loading and co-solvent type had no effect on Remdesivir's morphology. For excipients, high peaks of mannitol (13.5, 17, 18.5, 20.2, 21, 22, 24.5, 25, 27.5, and 36 degree two-theta) were detected in a number of crystallizations, showing that mannitol

Table 3 Relevant crystallographic data for **Remdesivir-I** and **II**.

Crystal data	Remdesivir-I	Remdesivir-II
Chemical formula	$C_{27}H_{35}N_6O_8P$	$C_{27}H_{35}N_6O_8P$
M_r	602.58	602.58
Crystal system, Space group	Triclinic, $P1$	Monoclinic, $P2_1$
Temperature (K)	170(2)	170(2)
a, b, c (Å)	8.5565(11), 10.5456 (16), 17.147(2)	10.5286(17), 12.809(2), 11.1106(19)
α, β, γ (°)	96.105(4), 99.219(4), 94.937(4)	90, 100.022(5), 90
Volume (Å^3)	1510.1(4)	1475.6(4)
$Z, D_c/(g/cm^{-3})$	2, 1.34139	2, 1.356
Radiation type	GaKα	MoKα
F(000)	636	636
Data collection		
T_{min}, T_{max}	0.5372, 0.7508	0.4799, 0.7455
μ (mm^{-1})	0.837	0.152
Measured, and independent	24511, 24511	28620, 6455
Observed reflections	9621	5990
Flack parameter	0.09(4)	$-0.01(4)$
R_{int}	0.1356	0.0606
Refinement		
$R[F^2 > 2\sigma(F^2)]$	0.0885	0.0572
$wR(F^2)$	0.2506	0.1538
S	0.998	1.077
No. of refined parameters	860	384
$\Delta\rho_{max}$ (e Å^{-3})	-0.339	-0.332
$\Delta\rho_{min}$ (e Å^{-3})	0.299	0.961

remained crystalline as a mixture of δ and α form in these formulations [17]. Similarly, certain leucine peaks (6 and 19 degree two-theta) were identified in all crystallized TFF Remdesivir–leucine formulations, demonstrating that leucine retained its crystallinity during the procedure. Captisol® and lactose, on the other hand, remained amorphous during the TFF procedure.

4. Thermal methods of analysis

4.1 Melting behavior

The melting enthalpy heat-flow curves for the Remdesivir at 127 and 137 °C, respectively, clearly demonstrate the crystalline character of the material [16,18].

4.2 Differential scanning calorimetry

Sahakijpijarn et al. [16] performed thermal analysis of powder samples using a differential scanning calorimeter Model Q20 fitted with a refrigerated cooling system (TA Instruments Inc., New Castle, DE, USA) (RCS40, TA Instruments Inc., New Castle, DE, USA). Samples weighing 2–3 mg were weighed and placed into a T-zero pan to be used later. Before placing the pan in the sample holder, the T-zero hermetic cover was crimped and a hole bored in the lid. To determine the glass transition temperature and glass-forming capabilities of unprocessed Remdesivir powder, samples were heated at a rate of 10 °C/min from 25 to 150 °C, cooled to 40 °C, and then heated to 250 °C. To determine the crystallinity of the TFF formulations, samples were heated from 25 to 350 degrees Celsius at a rate of 5 °C/min. The scans were conducted with a modulation period of 60 s and an amplitude modulation of 1 °C. Throughout the analysis, dry nitrogen gas at a flow rate of 50 mL/min was utilized to cleanse the DSC cell. TA Instruments Trios V.5.1.1.46572 software was used to process the data (TA Instruments, Inc., New Castle, DE, USA).

Modulated differential scanning calorimetry (mDSC) was used to measure Remdesivir's glass-forming ability and glass transition temperature (Tg) in each formulation. The mDSC thermogram of Remdesivir unprocessed powder is shown peak with enthalpy 103.56 J/g °C. The initial heating cycle established that Remdesivir was crystalline and possessed a melting point of 133 °C. The cooling and second heating cycles revealed that Remdesivir's glass transition temperature was around 60 °C. In no cycle was a recrystallization peak seen.

Yu et al. [15] performed thermal analysis on RDV-I and RDV-II samples using a differential scanning calorimeter (TA DSC Q100). A powder sample weighing approximately 2.4 mg was placed in an aluminum pan and heated at a rate of $10 °C min^{-1}$ with a nitrogen flow rate of 50 mL. min^{-1} over a temperature range of 20–200 °C. DSC analysis showed that RDV-II is a more thermodynamically stable polymorph.

4.3 Thermogravimetric analysis

Thermal Gravimetric Analysis (TGA) of Remdesivir was obtained using a Perkin Elmer pyris 1 apparatus. The sample (2.042 mg) were placed in Aluminum pan, pierced prior to scan, and temperature profile 50–550 °C at a rate of 10 °C/min under nitrogen purge (50 mL/min). Gravimetric Analysis (TGA) can be seen that the powder of Remdesivir is chemically unstable above the melting point temperatures and the compound was decomposing at 199 °C. as shown in Fig. 2.

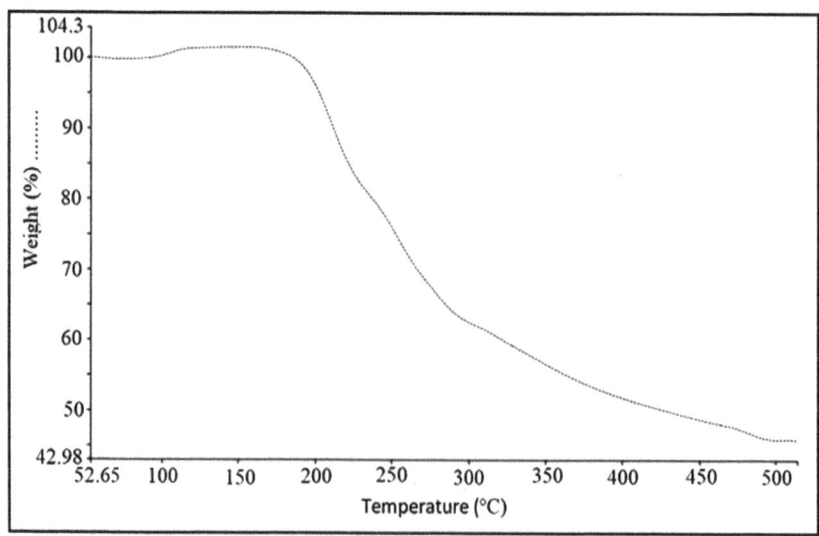

Fig. 2 Thermogravimetric analysis (TGA) of Remdesivir.

5. Characterization and identification

5.1 Spectroscopy

5.1.1 UV/Vis spectroscopy

A Shimadzu UV-spectrophotometer (model no. UV-1800) was used to record an ultraviolet absorption spectrum of Remdesivir (20 μg/mL) in methanol. The absorption spectra are measured in a 1 cm quartz cell within 200–400 nm range. The ultraviolet spectrum is displayed in Fig. 3 with two a maximum absorption of Remdesivir at 210 and 240 nm.

5.1.2 Spectrofluorometry

The fluorescence spectrum emission of Remdesivir in methanol was recorded using a Jasco FP-8200 Spectrofluorometry (Jasco Corporation, Japan) equipped with a 150 W xenon lamp and 1 cm quartz cells. The slit widths for both the excitation and emission monochromators were set at 5.0 nm. Remdesivir exhibited maximum excitation at 245 and maximum emission at 402 nm. The fluorescence spectrum of Remdesivir is shown in Fig. 4.

5.1.3 Infrared spectroscopy

The infrared absorption spectrum of Remdesivir was recorded as KBr disk using the Perkin Elmer FT-IR Spectrum BX apparatus. Fig. 5 showed the FT-IR of Remdesivir. The characteristic absorption bands are shown in Table 4.

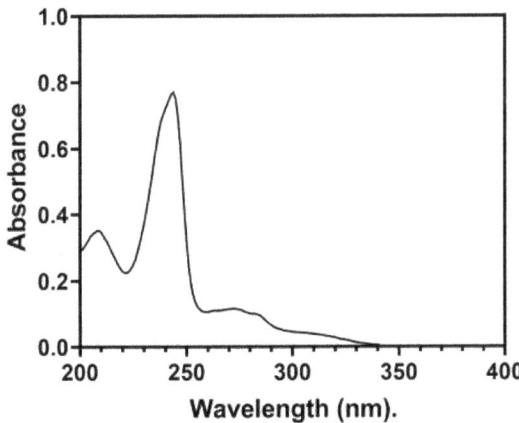

Fig. 3 The UV-absorption spectrum of 20 μg/mL of Remdesivir in methanol.

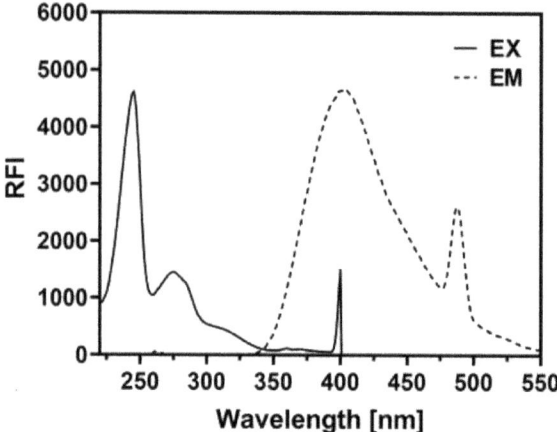

Fig. 4 The fluorescence spectrum excitation and emission of Remdesivir in methanol.

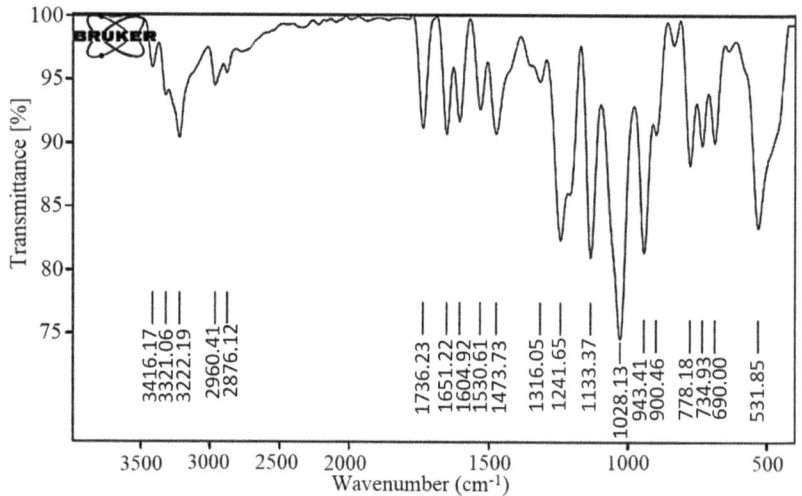

Fig. 5 IR spectrum of Remdesivir.

Table 4 Interpretation of Remdesivir FTIR spectra.

No.	Functional group	Standard value (cm^{-1})	Obtained value
1	C—O group	1050–1150	1133.37
2	C—N group	1000–1350	1241.65
3	C=C group	1600–1680	1604.92
4	C=O group	1640–1810	1736.23
5	C—H group	2850–3000	2960.41
6	O—H group	2500–3400	3222.19
7	N—H group	3300–3500	3416.17

5.1.4 Nuclear magnetic resonance spectrometry

5.1.4.1 ^{1}H NMR spectrum

^{1}H NMR spectrum of Remdesivir was scanned in DMSO-d6 on a Brucker NMR spectrometer operating at 500 MHz. Chemical shifts are expressed in δ-values (ppm) relative to TMS as an internal standard. Coupling constants (J) are expressed in Hz (Table 5 and Fig. 6).

5.2 Mass spectrometry

The mass spectrum of Remdesivir ($C_{27}H_{35}N_6O_8P$, 602.6) was obtained using an Agilent 6320 Ion trap mass spectrometer (Agilent technologies, USA) equipped with an electrospray ionization interface (ESI). A connector was used instead of a column. Mobile phase composed of a mixture of solvents A and B (50:50), where A is HPLC grade water, and B is acetonitrile. The compound was prepared by weighing the solid substances to $1 \, mg \, mL^{-1}$ in DMSO and diluted with mobile phase. The test solution was prepared by diluting the stock solutions to $10–30 \, mg \, mL^{-1}$ depending

Table 5 ^{1}H NMR of Remdesivir (DMSO-d6).

Signal	Location (δ)	Shape	Integration	Correspondences
1	7.93	s	2	$Ar—NH_2$
2	7.34	m	2	ArHs
3	7.18	m	3	ArHs
4	6.90	s	1	Pyrimidine—H
5	6.22	d	1	Pyrrole—Hs
6	6.34	d	1	Pyrrole—Hs
7	6.04	d	1	—OH
8	5.38	d	1	—OH
9	4.66	t	1	—NH—
10	4.25	m	3	Furan—Hs
11	4.09	t	2	$CH_2—O—C=O$
12	3.96	m	2	$CH_2—O—P=O$
13	3.86	m	1	—CH—C=O
14	1.42	m	1	$—CH—(CH_2—CH_3)_2$
15	1.26	m	7	$—(CH_2)_2—, —CH—CH_3$
16	0.78	td	6	$—CH—(CH_2—CH_3)_2$

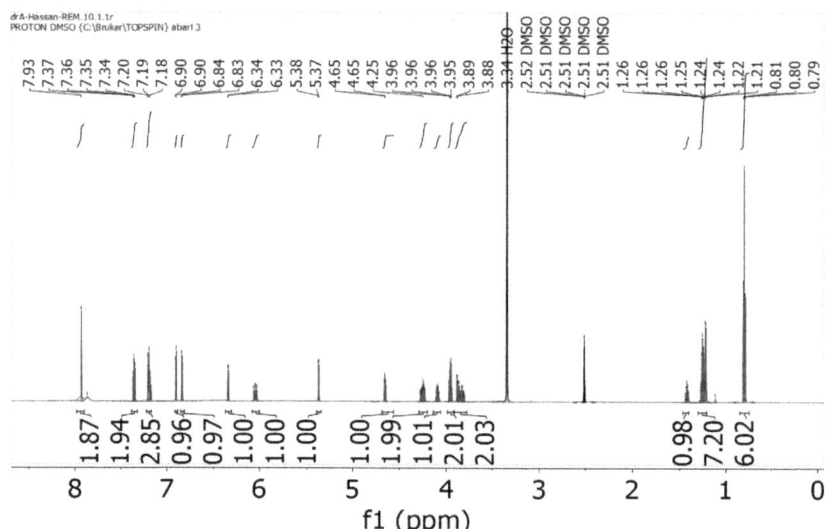

Fig. 6 ^{1}H NMR spectrum of Remdesivir.

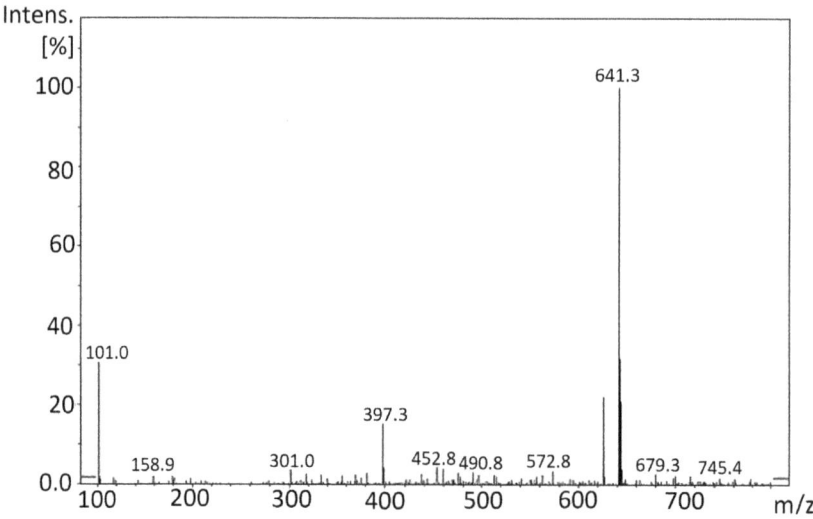

Fig. 7 Full scan mass spectra of Remdesivir.

on the ions intensities- with mobile phase. Flow rate was $0.4\,\text{mL}\,\text{min}^{-1}$, run time was 5 min. MS parameters were optimized for each compound. The scan was ultra-scan mode. MS2 scans were performed in the mass range of m/z 50–1000. The ESI was operated in positive mode. The source temperature was set to 350 °C nebulizer gas pressure of 55.00 psi; dry gas flow rate of 12.00 L min. Fig. 7 showed the peak at $m/z = 625.6\ [M+23]^+$

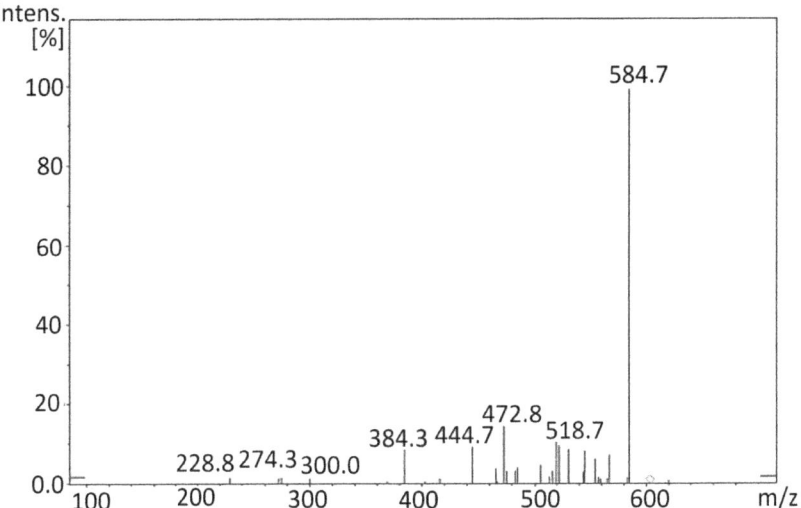

Fig. 8 Product ion spectra of Remdesivir.

due to $[M+Na]^+$ and the peak corresponding to $[M+K]^+$ appeared at 641.3 $[M+39]^+$. Fig. 8 showed the product ion spectra of Remdesivir.

6. Analytical profiles of drug substances and excipients

6.1 Compendial methods of analysis

The compendial monograph (pharmacopoeia) was not reported up to date.

6.2 Titrimetric methods of analysis

6.2.1 Non-aqueous titration

WHO propose a Remdesivir draft for inclusion for The International Pharmacopoeia [19]. They described a Non-aqueous titration method as assay of Remdesivir. Dissolving 0.4 g in 50 mL of glacial acetic acid and titrate with perchloric acid (0.1 mol/L). Each mL of perchloric acid (0.1 mol/L) is equivalent to 60.26 mg of Remdesivir.

6.2.2 Multiple isothermal titration calorimetry (ITC)

Researchers from the University of Spain and USA conducted multiple isothermal titration calorimetry (ITC) measurements and computational molecular dynamics simulations to perform a structural and thermodynamic study of the interaction between CAPTISOL and neutral or protonated Remdesivir. At pH 3, according to the results of the global analysis of the calorimetric measurements, a relatively strong complex with an association constant of 104 is generated. As a result of the potential of mean force (PMF) profiles, the values obtained for the association constants indicate that the binding is weaker for the structure with the lowest number of sulfobutylether (SBE) substitutions, which is consistent with the ITC conclusion that the affinity is significantly dependent on the number and location of structures containing CAPTISOL [20].

6.3 Electrochemical methods of analysis

The Remdesivir anti-COVID-19 medication electrochemical determination was analyzed theoretically for the first time. An anodic method utilizing the Squaring Dye—Ag_2O_2 combination was investigated in this study. The electroanalytical process's mechanism is branched, implying a rather dynamic nature. However, a mathematical model study based on linear stability theory and bifurcation analysis shows that the composite electroanalytical efficiency as an electrode modification is true, even though it is an electrode modification [18].

6.4 Spectroscopic methods of analysis

6.4.1 Spectrophotometry

Bulduk and Akbel [21] developed and verified UV spectrophotometric techniques for quantifying Remdesivir in pharmaceutical formulations. The UV spectra was recorded between 200 and 800 nm using deionized water as the solvent, and the wavelength of 247 nm was chosen. These techniques were verified in accordance with the protocols outlined in the ICH Q2 standards (R1). The approaches performed well in terms of linearity, accuracy, and recovery. Within a dosage range of 10–60 mg mL^{-1}, correlation values were better than 0.999.

6.4.2 Spectrofluorimetry

Elmansi et al. [22] developed and validate a spectrofluorimetric method for determination of Remdesivir in an Iv infusion and human plasma Measurement of Remdesivir was based on its natural fluorescence at pH 4 and wavelengths of 244/405 nm. Over the range of 1.0–65.0 ng/mL, calibration was accomplished. To obtain maximal sensitivity, different factors impacting the suggested approach were investigated, with detection and quantification limits of 0.287 and 0.871 ng/mL, respectively. The described approach is thought to be the first spectrofluorimetric method for Remdesivir estimation. The approach was also used to determine the drug's concentration in a designed IV infusion and spiked human plasma.

6.5 Chromatographic methods of analysis

6.5.1 Thin-layer chromatography-densitometric (TLC-densitometric)

Noureldeen et al. [23] proposed and validated a green densitometric method for determination of Remdesivir and Favipiravir in pharmaceutical formulations and spiked human plasma on normal phase TLC plates. The developing mobile phase system is ethyl acetate–methanolammonia (8:2:0.2 by volume). Remdesivir and Favipiravir have retardation factors (R_f) of 0.18 and 0.98, respectively. An ICH-approved validation study was conducted. They have outstanding sensitivity with quantitation limits of 0.12 and 0.07 μg/band. Applicable to tablet formulations and spiked plasma, the new approach yielded excellent recoveries between 97.21% and 101.31% the method's greenness was assessed using the Greenness Profile and EcoScale. There are four green profile quadrants and an Eco-Scale score of 80.

6.5.2 High performance liquid chromatography

In The WHO proposal monograph of Remdesivir [19], an HPLC method was described for the determination of Remdesivir and its related compounds in the finished product using a stainless steel column (4.6 mm × 25 cm) packed with end-capped particles of silica gel with chemically-bonded octadecylsilyl groups (5 μm). The mobile phase was a mixture of 35 volumes of phosphoric acid solution and 65 volumes of methanol and operate at a flow rate of 1.0 mL per minute, conducted with a UV spectrophotometer with a wavelength of 237 nm as a detector. A constant temperature of 25 °C was maintained in the column. For standard and test solutions, 5 μL each were injected and recorded in the chromatogram for 25 min. The percentage content of $C_{17}H_{35}N_6O_8P$ was calculated in the sample using the stated content of $C_{17}H_{35}N_6O_8P$ in Remdesivir and the areas of the peaks corresponding to Remdesivir obtained in the chromatograms of standard and test solutions.

In the DAICEL Chiral Technologies proposal analytical methods of Remdesivir, Umstead [24] described an HPLC method for chiral separation and determination of Remdesivir and several of its key-starting materials in final product. A CHIRALPAK® IA-3 (250 mm × 4.6 mm i.d.) column was used, with a mixture of (n-Hexane:Ethanol:IPA:Ethanolamine:Formic acid) (80:5:15:0.05:0.1) (v/v/v/v/v) as a mobile phase and operated at a flow rate of 1.5 mL/min. A UV spectrophotometer was utilized with a wavelength of 245 nm ref. 450 nm as a detection wavelength. A constant temperature of

40 °C was maintained in the column. A standard solution (1.0 mg/mL) was prepared in (Hexane:Ethanol) (50/50) (v/v) and 10 μL was injected each and the chromatogram was recorded for 40 min.

In the Waters proposal, Remdesivir's analytical methods were investigated for separation and determination of Remdesivir to monitor biotransformation events.

Alden et al. [25] described liquid chromatography paired with optical detection or with mass spectrometric method for nucleotide analogs separated using tributylamine as an ion pairing reagent. An Atlantis PREMIER BEH C18 AX 1.7 mm Column (2.1 × 50 mm) was used. The incremental gradient elution was used to separate mixed analytes with polarities that range across a wide range in a single run. Chromatographic mobile phases were produced on-line with IonHance buffer concentrates (which include 20% (v/v) acetonitrile) using a quaternary pump. The buffer concentrates were generated using a 1:5 dilution method to obtain final concentrations of 100 mM in 4% acetonitrile for the IonHance CX-MS Concentrate A, pH 5, and 200 mM in 4% acetonitrile for the IonHance Ammonium Acetate pH 6.8 Concentrate. To create the gradient, the 1:5 dilutions were combined with 18 M water and acetonitrile. The final gradient was 5 mM ammonium acetate (pH 6.8) in 0% acetonitrile to 20 mM ammonium acetate (pH 6.8) in 60% acetonitrile in 4 min using a linear gradient, followed by a 0.5-min return to the initial concentration. Additionally, a longer 8-min gradient was tested with favorable results. and run at 0.5 mL/min. Detector 1 is an Acquity Premier PDA Detector. Detector 2 is an ACQUITY QDa Mass Detector. Maintain the column and samples at 50 °C and 12 °C, respectively, at all times.

Reddy et al. [26] used high-performance liquid chromatography with an ultraviolet detector to develop an unique validated liquid chromatographic method for the quantitative measurement of degradation products in Remdesivir Injectable Pharmaceutical Products. An octyldecylsilane chemically bonded column (Kromasil KR100-5 C18; USP L1 phase) with dimensions of 250 mm length, 4.5 mm inner diameter, and 5 μm particle size was used to improve the procedure. The method was validated in accordance with current regulatory criteria for analytical method validation, including the International Conference on Harmonization. In concentrations ranging from quantification level to 200% of the specification level of specified and unknown degrading contaminants. The method was successfully used to investigate degradation products in Remdesivir Injectable medicinal products.

Patel et al. [27] created a reverse phase high performance liquid chromatographic (RP-HPLC) technique for separation and quantification of Remdesivir in its API form. The separation was performed using a C18 (250 mm × 4.6 mm, 5 µ) column with Buffer, pH 5.0: Acetonitrile (30:70) as the mobile phase at a flow rate of 1 mL/min and detection at 253 nm. The retention time of the drug was 4.402-min. The method's linearity, accuracy, and precision were found to be within the accepted range. Remdesivir has a linearity of 10–30 µg/mL. The drug was exposed to stress conditions of hydrolysis, oxidation, photolysis, and thermal degradation; significant deterioration was identified in alkaline degradation.

6.5.3 Ultra performance liquid chromatography-tandem mass spectrometer (UPLC-MS/MS)

After a one-step protein precipitation process, a HPLC-MS/MS technique was established to quantify active metabolite of Remdesivir (GS-441524 (Nuc)) in rat plasma samples by Du et al. [28]. Under gradient elution conditions, the mobile phase A (ACN:water ratio of 95:5, v/v, 0.1% formic acid) and phase B (water:ACN ratio of 99:1, v/v, 0.1% formic acid) were made fresh. To obtain the baseline separation, the following gradient elution protocol was used: 1.2 min, 99% B; 1.2–3.5 min, 5% B; 3.5–3.6 min, 99% B; 3.6–4.5 min, 99% B. To eliminate the possibility of carryover, methanol:water (1:1, v/v) was utilized. Chromatographic separation was achieved on a Waters XBrige C_{18} column (50 × 2.1 mm, 3.5 µm) with a 4.5-min running time. In electrospray positive ion mode, several reaction monitoring transitions for Nuc were m/z 292.2 → 163.2 and 237.1 → 194.1 for the internal standard (carbamazepine). Maintain a constant temperature of 40 °C in the column. For 1.0 mg/mL solution of GS-441524 (Nuc) solution was prepared in (MeOH:water) (1:1) (v/v), inject 1 µL each and record the chromatogram for 4.5 min. The IS was used to determine Nuc's linearity. The Nuc calibration curve was linear in a range of 2–1000 ng mL^{-1} and a coefficient correlation (r) of greater than 0.990.

Pasupuleti et al. [19] developed analytical method to monitor Remdesivir drug profile in human plasma for pharmacokinetics (PK) and therapeutic drug monitoring (TDM). For the quick detection of Remdesivir in human plasma, the authors present an improved vortex-assisted salt-induced liquid-liquid micro-extraction (VA-SI-LLME) technique in combination with UHPLC-PDA and UHPLC-MS/MS. This procedure entails a single step of precipitating proteins with hydrochloric acid and then extracting with acetonitrile for analysis. Under optimal conditions (500 L acetonitrile + 2.5 g ammonium sulfate vortex extraction for 2 min), the correlation coefficient of 0.9969 was obtained for UHPLC-PDA (measured at 254 nm) and 0.9990 for UHPLC-MS/MS (measured at electrospray ionization with + ion mode transitions of m/z $603.1 \rightarrow$ m/z 402.20 and m/z $603.1 \rightarrow$ m/z 199.90). under optimal VA-SI-LLME conditions. The detection and quantification limits for UHPLC/PDA were 1.5 and 5 ng/mL, and 0.3 and 1 ng/mL for UHPLC-MS/MS, respectively. The approach yielded extraction recoveries between 90.79–116.74% and 85.68–101.34% for UHPLC/PDA and UHPLC-MS/MS, respectively, with intraday and interday precision $\leq$9.59 for both methods.

Avataneo et al. [29] developed UHPLC-MS/MS determination of Remdesivir and GS-441524 in human plasma. Standards and quality controls for Remdesivir and GS-441524 were developed in plasma from healthy donors. Protein precipitation was used to prepare the sample, which was then diluted and injected into the QSight 220 UHPLC-MS/MS apparatus. Chromatographic separation was accomplished using an Acquity HSS T3 1.8 m, 2.1 mm, 50 mm column using a gradient of water and acetonitrile containing 0.05% formic acid. The approach was verified in accordance with EMA and FDA regulations. There was a perfect fit between calibration curves and regression models that were "linear through zero". A 1/x weighting factor was used to make sure that the results were accurate at low concentrations, as well. Calculation coefficients (r^2) were all above 0.998 in all of the calibration curves.

The concentrations of RDV, GS-704277, and GS-441524 in plasma were measured by Humeniuk et al. [30] utilizing a liquid chromatography tandem mass spectroscopy approach with multiple reaction monitoring and electrospray ionization in the positive mode (QPS; LLC) (Newark, DE, USA). Multiple reaction monitoring was used to quantify the transitions m/z $603.3 \rightarrow 402.2$ and m/z $606.3 \rightarrow 402.2$ for RDV and an isotopically labelled internal standard (GS829143), m/z $441.1 \rightarrow 150.1$ and m/z $444.1 \rightarrow 150.1$ for GS-704277 and an isotopically labelled internal standard (GS829466), and m/z $292.2 \rightarrow 202.2$ and m/z $295.2 \rightarrow 205.2$ for GS-441524. The bioanalytical approach was validated using calibrated concentration ranges of 4–4000 ng/mL for RDV, 2–2000 ng/mL for GS-704277, and 2–2000 ng/mL for GS-441524. Inter-assay precision, as measured by coefficient of variation, was 2.1–5.3% for Remdesivir, GS-704277, and GS-441524, and accuracy, as measured by inter-assay percent relative error, was 9.8 to 9.5% for RDV, GS-704277, and GS-441524. All samples were evaluated within the interval specified by data on frozen stability storage.

For the quantification of Remdesivir and its active metabolites GS-441524, Alvarez et al. [31] developed and validated a method that relied on liquid chromatography coupled to triple quadrupole mass spectrometry detection and used 50 μL of plasma to detect the presence of Remdesivir and its active metabolites. A straightforward protein precipitation was performed using 75 μL methanol containing the internal standard (IS) Remdesivir-13C6 and 5 μL 1 M ZnSO$_4$. Following separation on a Kinetex® 2.6 μm Polar C18 100A LC column (100 2.1 mm i.d.), both chemicals were identified using electrospray ionization in positive mode on a mass spectrometer. For Remdesivir, the ion transitions were m/z 603.3, →m/z 200.0, and m/z 229.0; for GS-441524, the ion transitions were m/z 292.2, →m/z 173.1, and m/z 147.1; and for Remdesivir-13C6, the ion transitions were m/z 609.3→ m/z 206.0. The calibration curves for Remdesivir were linear in the 1–5000 g/L range and for GS-441524 were linear in the 5–2500 μg/L range, with the limit of detection set at 0.5 and 2 μg/L and the limit of quantification set at 1 and 5 g/L, respectively. Precisions for Remdesivir at 2.5, 400, and 4000 μg/L and for GS-441524 at 12.5, 125, and 2000 μg/L were less than 14.7% and accuracy was in the [89.6–110.2%] range. There was a minor matrix effect identified, which was accounted for by IS. On NaF-plasma, Remdesivir and its metabolite were shown to be more stable. After 200 mg IV single injection, Remdesivir concentrations immediately fell with a half-life of less than 1 hour, but GS-441524 concentrations increased fast and gradually decreased until H24 with a half-life of about 12 h.

Hu et al. [32] developed a high-performance liquid chromatography-tandem mass spectrometric method in a positive electrospray ionization mode for separating the active metabolite Remdesivir nucleotide triphosphate (RTP) from its precursor Remdesivir nucleotide monophosphate (RMP) using a BioBasic AX column. Chromatographic retention was tuned stepwise using an anion exchange column and matrix stability was enhanced using 5,5′-dithiobis-(2-nitrobenzoic acid) and PhosSTOP EASYpack, and recovery was raised by dissociating tight protein binding using a 2% formic acid aqueous solution. The technique has a quantitation limit of 20 nM for RMP and 10 nM for RTP. Validation of the method indicated appropriate precision (RSD 11.9% for RMP, RSD 11.4% for RTP) and accuracy (93.6%–103% for RMP, 94.5%–107% for RTP).

Xiao et al. [33] established and validated LC-MS/MS techniques for the measurement of Remdesivir and its metabolites GS-441524 and GS-704277 in human plasma. These methods incorporated two critical characteristics to ensure their precision, accuracy, and robustness. Stability difficulties with the analytes were addressed by treating plasma samples with diluted formic acid (FA) and each analyte was injected separately using distinct ESI modes and organic gradients to maximize sensitivity and reduce carryover. Chromatographic separation was accomplished with a 3.4 min run time on an Acquity UPLC HSS T3 column (2.1 × 50 mm, 1.8 µm). For Remdesivir, GS-441524, and GS-704277, the calibration ranges were 4–4000, 2–2000, and 2–2000 ng/mL, respectively. Across three QC levels, the intraday and interday precision (percent CV) for all three analytes was less than 6.6%, and the accuracy was less than 11.5%. Remdesivir, GS-441524, and GS-704277 plasma can be stored for 392, 392, and 257 days at a temperature of 70 °C for a long time.

Reckers et al. [34] established and validated an LC–MS/MS technique for quantifying Remdesivir, its metabolite GS-441524, and dexamethasone. The approach was applied to 23 blood samples from seven individuals with severe COVID-19. The detection limits for Remdesivir, GS-441524, and dexamethasone were 0.0375 ng/mL, 0.375 ng/mL, and 3.75 ng/mL, respectively. The levels of Remdesivir, GS-441524, and dexamethasone showed modest intra-patient variability but high inter-patient variability. The considerable inter-patient variability emphasizes the significance of therapeutic medication monitoring and probable dosage adjustments in COVID-19 patients in order to achieve effectiveness.

7. Determination of the drug in body fluids and tissues

7.1 Chromatographic methods

Warren et al. [35] developed a LC–MS/MS method for determination of the metabolites (GS-5734) concentration in the intracellular. The LC–MS/MS analysis was carried out utilizing low-flow ion-pairing chromatography, as described in Ref [36]. Analytes were separated using a 50 mm × 2.5 m Luna $C_{18}(2)$ HST column (Phenomenex) linked to an LC-20ADXR ternary pump system (Shimadzu) and an HTS PAL auto-sampler (LEAP Technologies). A multi-stage linear gradient from 10% to 50% acetonitrile in a mobile phase containing 3 mM ammonium formate (pH 5.0) and 10 mM dimethylhexylamine was used to separate analytes over an 8-min period at a flow rate of 150 µL min^{-1}. The detection was carried out using a positive ion and multiple reaction monitoring mode on an API 4000 (Applied Biosystems) MS/MS. The intracellular metabolites alanine metabolite, Nuc, nucleoside monophosphate, nucleoside diphosphate, and nucleoside triphosphate were measured using seven-point standard curves ranging from 0.274 to 200 pmol (about 0.5 to 400 μM) in untreated cell extract. Adenosine nucleotide levels were also measured to ensure that no dephosphorylation occurred during sample collection and preparation. A Countess automatic cell counter (Invitrogen) was used to count the total number of cells in each sample for determination of the metabolites concentration in the intracellular.

For plasma analysis, Using 20 nM 5-(2-aminopropyl)indole as an internal standard, aliquots of 25 μL each plasma sample were treated with 100 μL of 90% methanol/acetonitrile mixture and 10% water. On the other hand, the Agilent Captiva 96 well 0.2 μm filter plate was used to filter the samples. Filtered samples were dried completely for 20 min and reconstituted with 1% acetonitrile, 99.9% water, and 0.01% formic acid. Using an HTC Pal auto-sampler, an aliquot of 10 μL was injected for LC-MS/MS. Waters Acquity extreme performance LC (Waters Corporation, Milford, MA, USA) was used to separate the samples on a Phenomenex Synergi Hydro-RP 30A column (75 × 2.0 mM, 4.0 μm) with an analytical flow rate of 0.26 mL min^{-1} and a gradient from Mobile Phase A to Mobile phase B containing 0.2% formic acid in 99% water and 1% acetonitrile over 4.5 min. An electrospray probe on a Waters Xevo TQ-S was employed for MS/MS analysis. Calibration curves with eight points covering over three orders of magnitude in concentration were used to determine plasma levels of GS-734, alanine metabolism product, and nucleoside metabolite (Nuc). To ensure accuracy and precision within20%, quality control samples were taken at the start and end of the run.

8. Stability

8.1 Stability indicating method

For the measurement of Umifenovir and Remdesivir in tablet dosage form, a simple stability indicating reverse-phase high performance liquid chromatographic (RP-HPLC) approach has been devised by Surabhi and Jain [17]. Using the reference solution, the chromatographic solution was optimized. The chromatographic method employed a Zorbax SB C18 column of dimensions 150 × 4.6 mm, 3.5 μm, employing isocratic elution with an acetonitrile and water mobile phase in a 50:50 ratios for the chromatographic separation, which was monitored at a wavelength 230 nm PDA detector with a flow rate of 1 mL/min. The duration time of the run was ten minutes. The developed technique was validated in accordance with the ICH recommendations. The calibration curves plotted were linear, with a regression coefficient of R^2 greater than 0.999, indicating that the linearity was within the limit. As part of technique validation, recovery, specificity, linearity, accuracy, robustness, and ruggedness were determined, and the findings were and found to be within acceptable limits. All of the degradation products produced under the stress conditions are effectively separated, and the peaks have been successfully resolved with an appropriate retention time.

Remdesivir stability-indicating study was developed by Hamdy et al. utilizing HPLC technique [37]. For a chromatographic separation, C_{18} column (250 mm, 5 mm, 5 nm) was used with diode array detection and fluorescence detection. Acetonitrile and distilled water (acidified with phosphoric acid, pH 4) in the ratio of 55:45 (v/v) were employed for isocratic elution. HPLC-diode array detection had a linearity range of 0.1–15 µg/mL, whereas fluorimetric detection had a linearity range of 0.05–15 µg/mL. Accelerated alkaline, acidic, neutral hydrolysis, oxidative, heat, and photolytic stress conditions are all listed by the International Conference on Harmonization as degrading Remdesivir. This study confirm that the parent molecule has been degraded. The proposed approaches were able to identify the intact drug with no overlapping peaks in any of the assumptions that they were based on. Drug stability is endangered by heat and basic hydrolytic stresses, as demonstrated by extensive degradation.

8.2 Solid-state stability

According to ICH requirements, two commercial and two pilot scale batches of Remdesivir were stored in the planned commercial packaging for up to 48 months at 30 °C/75°RH and for up to 6 months at 40 °C/75°RH. These included appearance, moisture, assay, and impurity content. For up to 48 months, there was no loss in assay or rise in impurity content. Some batches showed a slight increase in water content between the first and third months, although this remained stable for up to 48 months. An ICH Q1B Photostability Testing of New Drug Substances and Products was performed on one pilot batch. Appearance, assay, impurity, and water content were assessed. The results show Remdesivir is not photolabile.

Stress studies were conducted at −20 °C for one month, 50 °C/ambient humidity for two weeks, and 60 °C/ambient humidity for 1 week to assess shipping and handling conditions. The appearance, assay, impurity content, and water content of samples were all determined. Observed little to no change Remdesivir is therefore stable for 1 month at −20 °C, 2 weeks at 50 °C, and 1 week at 60 °C. [13]. Stability studies were performed according to current ICH guidelines; Sterility and bacterial endotoxin testing also were performed. Photostability studies are adequately conducted and show that the proposed finished product is photostable. Thus, a shelf-life of 12 months with a storage restriction "store at 2–8 °C" is acceptable based on provided data for Concentrate for Solution for Infusion and the proposed shelf-life of 3 years is accepted for the 100 mg powder for concentrate for solution for infusion [13].

8.3 Solution-phase stability

The prepared diluted solution of Remdesivir is stable for 24 h at room temperature (20–25 °C) or 48 h at refrigerated temperature (2–8 °C) [38].

9. Pharmaceutical applications

Remdesivir (Veklury) has been authorized by the US Food and Drug Administration for use in adult and pediatric patients 12 years of age and weighing at least 40 kg (approximately 88 pounds) for the treatment of COVID-19 that requires hospitalization [29]. In addition, it can be used for other viral infection like Ebola [39].

10. Pharmacology

10.1 Pharmacodynamic properties

Remdesivir is activated intracellularly to create GS-443902 (an adenosine triphosphate analog) which has broad-spectrum action and specifically inhibits viral RNA polymerases. Remdesivir is activated intracellularly via hydrolase cleavage by carboxylesterases, resulting in the formation of GS-704277, an intermediate metabolite. Following the breaking of the phosphoramidate bond, the nucleoside analogue monophosphate, GS-441524-MP, is formed, which is then phosphorylated to provide the pharmacologically active nucleoside triphosphate, GS-443902. When GS-441524-MP is dephosphorylated, the nucleoside analog, GS-441524, is formed, which is less effective in re-phosphorylation. Remdesivir's intracellular metabolic route is represented in Fig. 9. Remdesivir and its metabolites (GS-704277 and GS-441524) can be detected in plasma, but the active triphosphate GS-443902 can only be detected intracellularly, with PBMCs serving as a clinical surrogate cell type for measuring activation of the active triphosphate GS-443902 [30] (Table 6).

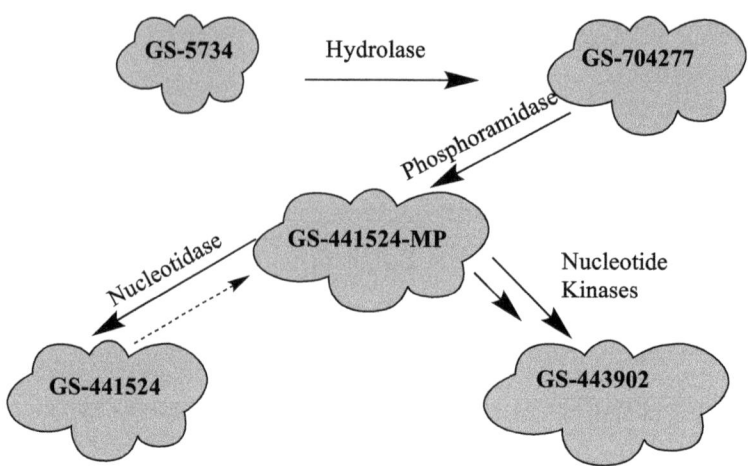

Fig. 9 Intracellular metabolic pathway of Remdesivir.

10.1.1 Mechanism of action

Remdesivir is a phosphoramide prodrug of the nucleoside adenosine C. When a medication is injected into the respiratory epithelial cells in the body, it is converted into its active form, nucleoside triphosphate, since it is a prodrug. The active form of Remdesivir can inhibit several coronaviruses from replicating in lung epithelial cells. SARS-CoV-2 is an enclosed, positive-sense, single-stranded RNA virus that is dominated by RNA-dependent RNA polymerase (RdRp), which is coded by the virus itself. After the virus infects the host cell, the viral genomic RNA is employed as a template, and RdRp is translated via the host cell protein synthesis pathway. RdRp is continually employed to finish the transcriptional synthesis of negative-strand subgenomic RNA, the production of numerous structural protein-related mRNAs, and viral genomic RNA replication. RdRp can accurately and effectively generate tens of thousands of nucleotides, allowing all other biological functions to occur after the virus has infected the host cell. The majority of anticoronavirus medicines that target RdRp are nucleoside analogues. The amount of viral RNA produced is reduced as a result. The prodrug is still being studied to see if it terminates RNA chains or induces mutations in the RNA. It was widely known that the prodrug Remdesivir hindered the function of RNA-dependent RNA polymerase, resulting in the elongation of the produced chain, based on Ebola virus investigations [40].

Table 6 The mean metabolites of Remdesivir [30].

Metabolites code	Metabolites structure
GS–5734	
GS–704277	
GS–441524–MP	
GS–441524	
GS–443902	

10.1.2 Clinical efficacy and safety

Angamo et al. [41] discovered that Remdesivir therapy resulted in a 21% improvement in clinical recovery rate on day 7 and a 29% rise on day 14. In addition, compared to the control group, Remdesivir-treated patients had a 39% lower risk of mortality on day 14; however, there was no meaningful difference on day 28. They also found that, despite a conditional recommendation against its use, Remdesivir was still effective in improving early clinical outcomes, reducing early mortality, and avoiding the use of high-flow supplemental oxygen and invasive mechanical ventilation in COVID-19 patients who were hospitalized. In comparison to placebo, Remdesivir was well tolerated with no notable Serious Adverse Effects (SAEs).

In another investigation, Lou et al. [42] found 5 RCT meta-analyses including 1782 individuals with COVID-19. Remdesivir outperformed the placebo-controlled group in terms of clinical improvement (relative risk (RR) = 1.17, 95% confidence interval (CI) = 1.07–1.29, $P = 0.0009$). The results of the Single-Arm Study Meta-analysis are listed below. During treatment of COVID-19 with Remdesivir, the pooled prevalence of clinical improvement meaningful results was 62%. During treatment of COVID-19 with Remdesivir, the incidence rates of acute renal damage, elevated hepatic enzymes, any significant adverse event, and death were 5% and 13%, respectively.

10.2 Pharmacokinetic properties

Deb et al. [43] discovered that the pharmacokinetic (PK) features of the parent Remdesivir, nucleoside core (GS-441524), and alanine metabolite (GS-704277) have been investigated in a limited number of clinical investigations. One of the trials used two highly infected COVID-19 patients in a PK study of the parent Remdesivir and nucleoside core, providing a 200 mg/day loading dose followed by 100 mg/day for the next 12 days. Similarly, pharmacokinetic characteristics were reported in a research with healthy volunteers who took 200 mg of Remdesivir per day for the first 4 days, then 100 mg per day for the next 4 days. Remdesivir and GS-441524 attained maximal concentrations in both experiments within 1 h. A peak serum concentration of 3027 ng/mL was attained immediately after intravenous (IV) infusion, followed by a rapid decline in plasma concentrations after 1 h. The parent Remdesivir's area under the curve (AUC) in 24 h ranged from 2.9 to 4.0 μg h/mL. The pharmacokinetic parameters of GS-441524 differed from those of Remdesivir. Patient 1 and patient 2 had plasma concentrations of 214 ng/mL, 316 ng/mL, and 206 ng/mL, and 113 ng/mL, 184 ng/mL, and 92.6 ng/mL, respectively, 1 h and 4 h after infusion. The AUC of the nucleoside core (GS-441524) has been found to range from 3.11 to 6.13 μg h/mL after 24 h. On day 1 and day 5, the Remdesivir C_{max} was 5.44 μg/mL and 2.61 μg/mL, respectively, whereas the GS-441524 C_{max} was 0.15 μg/mL and 0.14 μg/mL, respectively. On days 1 and 5, the AUC of parent Remdesivir was 2.92 and 1.56 μg h/mL, respectively, and 2.24 and 2.23 μg h/mL for GS-441524. Single or multiple dosage regimens of Remdesivir were tested for up to 14 days in a complete dose escalation trial with healthy participants. Remdesivir (3 mg to 225 mg) was administered as a 2-h IV infusion with a linear kinetics profile. After a single dosage of 3 mg to 225 mg Remdesivir, the C_{max} varied from 57.5 ng/mL to 4420 ng/mL, with each dose requiring 2 h to reach the peak plasma concentration. The AUC was also varied, ranging from 67.1 ng/mL to 5260 ng/mL. The clearance and volume of distribution, respectively, varied from 755 to 719 mL/min and 45.1 to 66.5 L. C_{max} and AUC showed similar dose-dependent patterns for GS-441524 and GS-704277 as well. In comparison to in vitro EC50 values found in SARS-CoV-2 experimental models, active triphosphate cellular concentrations were much greater (220–370-fold) following 75 mg or 150 mg single IV administration. Interestingly, the PK parameters of the parent Remdesivir prodrug at 150 mg/day for 7 or 14 days were comparable. However, between day 1 and day 7 or 14, the GS-441524 and GS-704277 were slightly different. Following daily dosing, the alanine metabolite raised by around 1.9-fold.

With just a few studies of Remdesivir PK in COVID-19 patients with renal failure, the present understanding of Remdesivir in particular populations such as renal or hepatic impairment is quite limited. In silico modeling of Remdesivir, pharmacokinetics suggests that its distribution is influenced by age, weight, liver, and renal function status. Because Remdesivir and its active metabolite are eliminated in the urine, dosage assessment data on individuals with renal impairment is required. Remdesivir plasma concentrations are higher in people with chronic renal disease. Healthy volunteers and renally compromised patients had 2.5- to 4-fold differences in critical PK characteristics of parent Remdesivir, such as AUC0-infinity, C_{max}, clearance, and volume distribution. Remdesivir should not be administered to individuals with an eGFR of less than 30 mL/min, according to the FDA. However, acute kidney damage was one of the most common side events in two clinical studies with Remdesivir. The solubility enhancer sulfobutylether-cyclodextrin sodium (SBECD) in Remdesivir's formulations may be to blame for the drug's renal impairment. SBECD is present in a variety of renally toxic medications (e.g., the IV formulation of voriconazole) and is excreted by the kidneys, with buildup occurring at CrCl 50 mL/min. SBECD, despite its buildup in renally compromised individuals, does not appear to induce significant renal damage, according to research. When using Remdesivir, it's still a good idea to keep an eye on your kidney function. Remdesivir, it turns out, can reduce the inflammatory response in mice with non-alcoholic fatty liver disease (NAFLD) caused by a high-fat diet. The primary theme of Remdesivir-mediated blockage of NAFLD development has been identified as the reduction of stimulator of interferon genes (STING), an endoplasmic reticulum membrane protein implicated in innate immune response. This raises the possibility that obese people with COVID-19 and liver disease may benefit from Remdesivir treatment in two ways, potentially affecting Remdesivir disposition, therapy, and safety.

10.3 Dosing

10.3.1 Adults and adolescents (≥12 years of age and weighing ≥40 kg)

Patients admitted to the hospital should receive a 200 mg intravenously infusion once, followed by a 100 mg intravenously infusion once-daily for 4 days (duration can be prolonged for an additional 5 days if there is no clinical response), or 10 days if the patient requires mechanical breathing and/or ECMO [44].

10.3.2 Pediatrics (<12 years of age or weighing <40 kg)

For patients of the pediatric weighing between 3.5 kg and 40 kg, the suggested dosage is a single loading dose of VEKLURY 5 mg/kg on Day 1, followed by VEKLURY 2.5 mg/kg once day starting on Day 2. For pediatric patients younger than 12 years of age and weighing at least 40 kg, a single loading dose of 200 mg on Day 1 is suggested, followed by once-daily maintenance doses of 100 mg starting on Day 2.

10.3.3 Geriatrics (>65 years of age)

Patients above the age of 65 years do not require any dose adjustments when using VEKLURY [45,46].

10.3.4 For renal impairment

• Remdesivir has not been subjected to a rigorous study in patients suffering from renal impairment [44].
• In patients with poor renal function (eGFR 30 mL/min), Remdesivir is not indicated due to the possibility of accumulation of sulfobutylether-cyclodextrin (SBECD), the excipient in the medication (this is the same issue with voriconazole IV). Nevertheless, many experts (including the

JH ABX Guide) support use due to the absence of significant toxicity when administered (voriconazole or Remdesivir) [44,46].

10.3.5 Hepatic impairment
VEKLURY's pharmacokinetics in patients with hepatic impairment have not been studied. There is no dose guideline for patients with hepatic impairment. VEKLURY should not be initiated in patients with an ALT level greater than five times the upper limit of normal [45,46].

10.4 ADME profile
ADME: IV Administration

10.4.1 Absorption
According to Deb et al. [43] study, Remdesivir comes in two different forms: a solution and a lyophilized powder. The solution is supplied in a 100 mg/20 mL vial with a 5 mg/mL concentration, and the lyophilized powder is packaged in a single-dose vial with a concentration of 100 mg. A 0.9% saline or a 5% glucose solution must be used to reconstitute the powder formulation. Intravenous infusion over 30 to 120 min is the most effective form of delivery. Remdesivir is given to adults in the following way: On day one, a 200 mg intravenous loading dose was given, followed by daily IV doses of 100 mg for 2–5 days or up to 10 days. Because intravenous administration allows for 100% absorption of Remdesivir, it is the recommended method. Because of the high hydrolysis-mediated first-pass clearance in the gastrointestinal tract, oral administration is not recommended, resulting in reduced absorption and systemic concentration. Remdesivir has been developed as a prodrug to boost cellular Remdesivir triphosphate concentrations. Intramuscular injection, while better than oral, would still result in subtherapeutic levels since Remdesivir is released slowly from the muscles, leading blood concentrations to fall below therapeutic levels. Since SARS-CoV-2 first targets the lungs, the inhaled route of administration of Remdesivir is now being investigated, and it is assumed that absorption would be sufficient to transport the medicine directly into the primary infected region. The inhaled formulation is primarily intended for outpatient usage and for COVID-19 individuals with less severe symptoms. The Remdesivir prodrug diffuses past the cellular membrane after gradual infusion and undergoes a series of hydrolysis events in the cytoplasm. When the nucleoside core (GS-441524) is hydrolyzed, a more water-soluble monophosphate form is formed, which is unable to go back and hence stays in the infected cells. As a result, when given as a slow intravenous infusion,

the ester prodrug of Remdesivir is well absorbed. Transporters can affect Remdesivir absorption in addition to the dose form and method of administration. P-glycoprotein (P-gp) and organic anion transporting polypeptides 1B1 (OATP1B1) transporters are substrates for Remdesivir. Similarly, Remdesivir's alanine metabolite is an OATP1B1 and OATP1B3 substrate. P-gp can efflux Remdesivir out of the cellular compartment depending on the transporter expression profile in the cell membrane, resulting in reduced intracellular levels and efficacy.

10.4.2 Distribution

According to Deb et al. [43] study, After intravenous administration of Remdesivir, the medication enters the tissues and blood cells via passive diffusion. The parent Remdesivir binds to human plasma proteins with an affinity of 88.0–93.6%, whereas GS-441524 and GS-704277 attach to proteins with 2% and 1%, respectively. Because of its instability within the tissues, Remdesivir is predicted to have a poor tissue distribution; yet, once its active metabolite penetrates the cells, it accumulates more than the extracellular prodrug form. With a single dosage of 10 mg to 225 mg, Remdesivir exhibited a volume of distribution of roughly 45.1–73.4 L in a cohort of eight participants. Similarly, a 7-day or 14-day multiple-dose (150 mg/day) research revealed a volume of distribution of 85.5 L. GS-441524 (nucleoside core) and GS-443902 (active triphosphate) concentrations in peripheral blood mononuclear cells (PBMCs) were utilized as markers of Remdesivir distribution throughout the cellular compartment. It's worth noting that intracellular concentrations were 220 to 370 times greater than the SARS-CoV-2 EC50. There is modest animal evidence of Remdesivir distribution to most tissues in rats and monkeys, but little or no evidence of Remdesivir presence in the brain, indicating that it does not pass the blood–brain barrier well. To attain larger concentrations in the affected organs, higher or more frequent dosage may be required. Remdesivir dosages more than 200 mg, on the other hand, have been connected to hepatotoxicity and renal failure. Thus, additional development of an inhalation dosage form is required so that Remdesivir can be delivered to infected cells at optimal concentrations while minimizing systemic toxicity and adverse medication responses.

10.4.3 Metabolism

In Deb et al. [43] study, Remdesivir is a prodrug that is converted to its active triphosphate form by a hydrolysis process (GS-443902). The prodrug is metabolized to the degree stated in parenthesis by carboxylesterase 1 (CES1) (80%), cathepsin A (10%), and CYP3A (10%). CYP2C8 and CYP2D6 were also discovered to have a role in the metabolism of Remdesivir in vitro. The Histidine Triad Nucleotide-binding Protein 1 metabolizes GS-704277 (HINT-1). The majority of the metabolites (74%) are identified in urine, while roughly 18% are found in feces. The nucleoside core metabolite (GS-441524) was the most common Remdesivir found in urine, with only 10% of the parent Remdesivir prodrug recovered. Hepatic blood flow, rather than metabolic enzymes, drives Remdesivir clearance; this is thought to be true due to Remdesivir's high extraction ratio and short elimination half-life.

10.4.4 Elimination

According to Deb et al. [43] study, Remdesivir's nucleoside core (GS-441524) is the most prevalent derivative found in urine, with Remdesivir and other metabolites present in moderate to low amounts. Parent Remdesivir and GS-704277 (alanine metabolite) are excreted mostly by biotransformation, but GS-441524 is removed primarily through renal processes such as glomerular filtration and active tubular secretion. Approximately 10% of the parent Remdesivir is excreted as a metabolite in the urine. In contrast, 49% of the dosage is excreted in the urine as GS-441524. In the feces, no Remdesivir or alanine metabolites were found, and only 1% of the dosage was eliminated as the nucleoside core. The half-life of GS-441524 is 27 h, compared to 1 h and 1.3 h for Remdesivir and GS-704277, respectively. The plasma half-life is around 1 h, whereas the intracellular nucleoside triphosphate half-life is 14 to 24 h. Because of its anionic charge, the nucleoside triphosphate stays in the cellular compartment for a longer time. Remdesivir's nucleoside triphosphate nucleoside half-life allows for once-daily dosage. Patients with renal impairment or those receiving renal replacement therapy such as dialysis or hemodialysis should avoid Remdesivir since it is mostly eliminated by the kidneys. Patients with an eGFR of more than 30 mL/min can take Remdesivir without needing to modify their dosage, but once the eGFR falls below 30 mL/min, Remdesivir is no longer indicated.

Acknowledgments

The authors wish to express their gratitude and thanks to the Pharmaceutical Chemistry and Central Laboratory in the College of Pharmacy at King Saud University for their analysis of samples with different techniques and assistance during this chapter.

References

[1] J.J. Malin, I. Suárez, V. Priesner, G. Fätkenheuer, J. Rybniker, Remdesivir against COVID-19 and other viral diseases, Clin. Microbiol. Rev. 34 (1) (2020). p. e00162-20.

[2] Y. Pashaei, Analytical methods for the determination of remdesivir as a promising antiviral candidate drug for the COVID-19 pandemic, Drug Discov. Ther. 14 (6) (2021) 273–281.

[3] Remdesivir (Compound), 2022. [cited 2022 24 March]; from https://pubchem.ncbi.nlm.nih.gov/compound/121304016#section=3D-Conformer.

[4] Remdesivir (HMDB0304869), The Metabolomics Innovation Centre (TMIC), 2022, [cited 2022 24 March]; from https://hmdb.ca/metabolites/HMDB0304869.

[5] Remdesivir, The Merck Index Online 2022, 2022, [cited 2022 24 March]; from https://www.rsc.org/Merck-Index/monograph/m12252/remdesivir?q=unauthorize.

[6] Remdesivir, 2022. [cited 2022 24 March]; from https://www.simsonpharma.com/product/2-ethylbutyl-r-2r-3s-4r-5r-5-4-aminopyrrolo-2-1-f-1-2-4-triazin-7-yl-5-cyano-3-4-dihydroxytetrahydrofuran-2-yl-methoxy-phenoxy-phosphoryl-l-alaninate.

[7] 3QKI37EEHE, 2022. [cited 2022 24 March]; from https://pubchem.ncbi.nlm.nih.gov/substance/328083432.

[8] Remdesivir, ChemSpider Search and Share Chemistry 2022 2022, [cited 2022 24 March]; from http://www.chemspider.com/Chemical-Structure.58827832.html?rid=afcba29c-0108-40a8-a23b-68f20f7bfb2c&page_num=0.

[9] Remdesivir, Drugs.com updated 2 Mar 2022; from, 2022. https://www.drugs.com/mtm/remdesivir.html.

[10] D. Siegel, H.C. Hui, E. Doerffler, M.O. Clarke, K. Chun, L. Zhang, S. Neville, E. Carra, W. Lew, B. Ross, Q. Wang, L. Wolfe, R. Jordan, V. Soloveva, J. Knox, J. Perry, M. Perron, K.M. Stray, O. Barauskas, J.Y. Feng, Y. Xu, G. Lee, A.L. Rheingold, A.S. Ray, R. Bannister, R. Strickley, S. Swaminathan, W.A. Lee, S. Bavari, T. Cihlar, M.K. Lo, T.K. Warren, R.L. Mackman, Discovery and synthesis of a phosphoramidate prodrug of a pyrrolo[2,1-f][triazin-4-amino] adenine C-nucleoside (GS-5734) for the treatment of Ebola and emerging viruses, J. Med. Chem. 60 (5) (2017) 1648–1661.

[11] T. Vieira, A.C. Stevens, A. Chtchemelinine, D. Gao, P. Badalov, L. Heumann, Development of a large-scale cyanation process using continuous flow chemistry en route to the synthesis of remdesivir, Org. Proc. Res. Dev. 24 (10) (2020) 2113–2121.

[12] K. Kumar Palli, P. Ghosh, S. Krishna Avula, B.S.S. Rao, A.D. Patil, S. Ghosh, G. Sudhakar, C.R. Reddy, P.S. Mainkar, S. Chandrasekhar, Total synthesis of remdesivir, Tetrahedron Lett. (2021), 153590.

[13] EMEA, Assessment report—Veklury, Procedure No. EMEA/H/C/005622/0000, 2020. 25 June 2020; from https://www.ema.europa.eu/en/documents/assessment-report/veklury-epar-public-assessment-report_en.pdf Retrieved October 23, 2021.

[14] S. Sekharan, X. Liu, Z. Yang, X. Liu, L. Deng, S. Ruan, Y. Abramov, G. Sun, S. Li, T. Zhou, B. Shi, Q. Zeng, Q. Zeng, C. Chang, Y. Jin, X. Shi, Selecting a stable solid form of remdesivir using microcrystal electron diffraction and crystal structure prediction, RSC Advances 11 (28) (2021) 17408–17412.

[15] K. Yu, S. Chen, C. Amgoth, G. Tang, H. Bai, X. Hu, Two polymorphs of Remdesivir: Crystal Structure, solubility, and pharmacokinetic study, CrystEngComm 23 (2021).

[16] S. Sahakijpijarn, C. Moon, J.J. Koleng, D.J. Christensen, R.O. Williams Iii, Development of remdesivir as a dry powder for inhalation by thin film freezing, Pharmaceutics 12 (11) (2020) 1002.

[17] S.R. Surabhi, N. Jain, Validated stability indicating method for determination of umifenovir-remdesivir in presence of its degradation products, Int. J. Dev. Res. 11 (2021) 6.

[18] V.V. Tkach, M. Kushnir, S.C. de Oliveira, J. Ivanushko, A.V. Velyka, A.F. Molodianu, P.I. Yagodynets, Z.O. Kormosh, L. Vaz dos Reis, O.V. Luganska, Theoretical description for anti-COVID-19 drug Remdesivir electrochemical determination, assisted by squaraine Dye–Ag_2O_2 composite, Biointerface Res. Appl. Chem. 11 (2) (2021) 9201–9208.

[19] Organization, W.H, WHO Drug Information 2020, vol. 34, WHO Drug Information, 2021, pp. 862–876. 4 [full issue].

[20] Á. Piñeiro, J. Pipkin, V. Antle, R. Garcia-Fandino, Remdesivir interactions with sulphobutylether-β-cyclodextrins: a case study using selected substitution patterns, J. Mol. Liquids 346 (2022) 117157.

[21] I. Bulduk, E. Akbel, A comparative study of HPLC and UV spectrophotometric methods for remdesivir quantification in pharmaceutical formulations, J. Taibah Univ. Sci. 15 (1) (2021) 507–513.

[22] H. Elmansi, A.E. Ibrahim, I.E. Mikhail, F. Belal, Green and sensitive spectrofluorimetric determination of Remdesivir, an FDA approved SARS-CoV-2 candidate antiviral; application in pharmaceutical dosage forms and spiked human plasma, Anal. Meth. 13 (23) (2021) 2596–2602.

[23] D.A. Noureldeen, J.M. Boushra, A.S. Lashien, A.F.A. Hakiem, T.Z. Attia, Novel environment friendly TLC-densitometric method for the determination of anti-coronavirus drugs "Remdesivir and Favipiravir": green assessment with application to pharmaceutical formulations and human plasma, Microchem. J. 174 (2022) 107101.

[24] W.J. Umstead, The chiral separation of remdesivir and several of its key starting materials, LC-GC North America 39 (6) (2021) 291–293.

[25] B.A. Alden, P. Christensen, D. Foley, L.J. Calton, S. Barnes, G. Gallo, M.A. Lauber, Comprehending COVID-19: Mixed-Mode Chromatography for Ion Pairing Free LC-MS of Remdesivir and Remdesivir Triphosphate, Waters Corporation, 2021.

[26] H.R. Reddy, S.R. Pratap, N. Chandrasekhar, S.Z.M. Shamshuddin, A novel liquid chromatographic method for the quantitative determination of degradation products in remdesivir injectable drug product, J. Chromat. Sci. (2021).

[27] V. Patel, N. Tiwari, K. Patel, Stability indicating RP-HPLC method development and validation for the estimation of Remdesivir in API form, World J. Pharm. Pharmaceut. Sci. 10 (6) (2021) 1544–1551.

[28] P. Du, G. Wang, S. Yang, P. Li, L. Liu, Quantitative HPLC-MS/MS determination of Nuc, the active metabolite of remdesivir, and its pharmacokinetics in rat, Anal. Bioanal. Chem. 413 (23) (2021) 5811–5820.

[29] V. Avataneo, A. de Nicolò, J. Cusato, M. Antonucci, A. Manca, A. Palermiti, C. Waitt, S. Walimbwa, M. Lamorde, G. di Perri, A. D'Avolio, Development and validation of a UHPLC-MS/MS method for quantification of the prodrug remdesivir and its metabolite GS-441524: a tool for clinical pharmacokinetics of SARS-CoV-2/COVID-19 and Ebola virus disease, J. Antimicrob. Chemother. 75 (7) (2020) 1772–1777.

[30] R. Humeniuk, A. Mathias, B.J. Kirby, J.D. Lutz, H. Cao, A. Osinusi, D. Babusis, D. Porter, X. Wei, J. Ling, Pharmacokinetic, pharmacodynamic, and drug-interaction profile of remdesivir, a SARS-CoV-2 replication inhibitor, Clin. Pharmacokinet. (2021) 1–15.

[31] J.-C. Alvarez, P. Moine, I. Etting, D. Annane, I.A. Larabi, Quantification of plasma remdesivir and its metabolite GS-441524 using liquid chromatography coupled to tandem mass spectrometry. Application to a covid-19 treated patient, Clin. Chem. Lab. Med. 58 (9) (2020) 1461–1468.

[32] W. Hu, L. Chang, C. Ke, Y. Xie, J. Shen, B. Tan, J. Liu, Challenges and stepwise fit-for-purpose optimization for bioanalyses of remdesivir metabolites nucleotide monophosphate and triphosphate in mouse tissues using LC–MS/MS, J. Pharmaceut. Biomed. Anal. 194 (2021) 113806.

[33] D. Xiao, K.H. John Ling, T. Tarnowski, R. Humeniuk, P. German, A. Mathias, J. Chu, Y.-S. Chen, E. van Ingen, Validation of LC-MS/MS methods for determination of remdesivir and its metabolites GS-441524 and GS-704277 in acidified human plasma and their application in COVID-19 related clinical studies, Anal. Biochem. 617 (2021) 114118.

[34] S.V. Gandhi, B.G. Kapoor, Development and validation of UV spectroscopic method for estimation of baricitinib, J. Drug Deliv. Ther. 9 (4-s) (2019) 488–491.

[35] T.K. Warren, R. Jordan, M.K. Lo, A.S. Ray, R.L. Mackman, V. Soloveva, D. Siegel, M. Perron, R. Bannister, H.C. Hui, Therapeutic efficacy of the small molecule GS-5734 against Ebola virus in rhesus monkeys, Nature 531 (7594) (2016) 381–385.

[36] L. Durand-Gasselin, K.K. Van Rompay, J.E. Vela, I.N. Henne, W.A. Lee, G.R. Rhodes, A.S. Ray, Nucleotide analogue prodrug tenofovir disoproxil enhances lymphoid cell loading following oral administration in monkeys, Mol. Pharm. 6 (4) (2009) 1145–1151.

[37] M.M.A. Hamdy, M.M. Abdel Moneim, M.F. Kamal, Accelerated stability study of the ester prodrug remdesivir: recently FDA-approved Covid-19 antiviral using reversed-phase-HPLC with fluorimetric and diode array detection, Biomed. Chromatogr. 5 (10) (2021).

[38] FDA, Fact sheet for healthcare providers emergency use authorization (EUA) OF VEKLURY® (remdesivir) for hospitalized pediatric patients weighing 3.5 kg to less than 40 kg or hospitalized pediatric patients less than 12 years of age weighing at least 3.5 kg. 2020, 2021, Available from: Retrieved October 26, 2022 [cited 2022 24 March]. from https://www.fda.gov/media/137566/download.

[39] EMA, Annex I—Conditions of use, conditions for distribution and patients targeted adressed to member states (Remdesivir gilead), 2020. from https://www.ema.europa. eu/en/documents/other/conditions-use-conditions-distribution-patients-targeted-conditions-safety-monitoring-addressed_en-3.pdf Retrieved October 25, 2021.

[40] I. Kumar, A review on pharmacokinetics, pharmacodynamics and clinical aspects of remdesivir and favipiravir for the treatment of coronavirus disease, J. Drug Deliv. Ther. 11 (1) (2021) 121–129.

[41] M.T. Angamo, M.A. Mohammed, G.M. Peterson, Efficacy and safety of remdesivir in hospitalised COVID-19 patients: a systematic review and meta-analysis, Infection (2021).

[42] L. Lou, H. Zhang, Z. Li, B. Tang, Z. Li, The efficacy and safety of remdesivir in the treatment of patients with COVID-19: a systematic review and meta-analysis, medRxiv (2021). p. 2021.03.12.21253470.

[43] S. Deb, A.A. Reeves, R. Hopefl, R. Bejusca, ADME and pharmacokinetic properties of remdesivir: its drug interaction potential, Pharmaceuticals 14 (7) (2021) 655.

[44] K.P.D.B. Dzintars, E.P.D. Avdic, Remdesivir, 2022.

[45] Gilead Sciences Canada Inc., Product monograph: remdesivir for injection; remdesivir solution for injection, 2022, [cited 2022 24 March]; from https://pdf.hres.ca/dpd_pm/00057134.PDF.

[46] Gilead Sciences Ireland UC, Veklury (Remdesivir): EU Summary of Product Characteristics, 2022, [cited 2022 24 March]; from https://www.ema.europa.eu/en/documents/other/veklury-product-information-approved-chmp-25-june-2020-pending-endorsement-european-commission_en.pdf.

CHAPTER FOUR

Vandetanib

Ahmed I. Al-Ghusn[a], Ahmed H. Bakheit[a,b], Mohamed W. Attwa[a], and Haitham AlRabiah[a]

[a]Department of Pharmaceutical Chemistry, College of Pharmacy, King Saud University, Riyadh, Kingdom of Saudi Arabia
[b]Department of Chemistry, Faculty of Science and Technology, Al-Neelain University, Khartoum, Sudan

Contents

Profiles of Drug Substances, Excipients, and Related Methodology, Volume 48
ISSN 1871-5125
https://doi.org/10.1016/bs.podrm.2022.11.004

1. Description

Vandetanib (Caprelsa®) is a 4-anilinoquinazoline developed by AstraZeneca as an antineoplastic kinase inhibitor of tumor angiogenesis and tumor cell proliferation and it can be used for other types of tumor. On April 6, 2011, the FDA authorized vandetanib to treat adults with non-resectable, locally progressed, or metastatic medullary thyroid carcinoma [1].

1.1 Nomenclature [2]

1.1.1 IUPAC name

- N-(4-bromo-2-fluorophenyl)-6-methoxy-7-[(1-methylpiperidin-4-yl) methoxy] quinazolin-4-amine
- N-(4-Bromo-2-fluorophenyl)-6-methoxy-7-((1-methyl-4-piperidinyl) methoxy)-4-quinazolinamine
- 4-(4-Bromo-2-fluoroanilino)-6-methoxy-7-(1-methylpiperidin-4-yl) methoxyquinazoline

1.1.2 Nonproprietary name

Generic: Vandetanib.
Synonyms: ZD6474, UNII-YO460OQ37K.

1.1.3 Proprietary names

Brand Names: Caprelsa, Zactima.

1.2 Formulae [2]

1.2.1 Empirical formula, Molecular weight, and CAS no

The empirical formula, molecular weight, and CAS number of the Vandetanib were displayed in Table 1.

1.2.2 Structural formula

Vandetanib's two-dimensional structural formula was displayed in Fig. 1.

1.3 Elemental analysis

See Table 2.

Table 1 Empirical formula, molecular weight, and CAS number.

Empirical Formula	$C_{22}H_{24}BrFN_4O_2$
Molecular Weight	475.4
CAS Number	443913-73-3

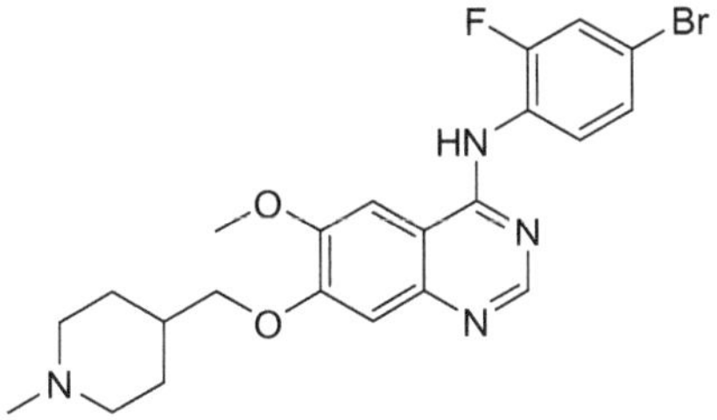

Fig. 1 Chemical structure of Vandetanib.

Table 2 The theoretical elemental
analysis and composition of Vandetanib
were detailed in Table 2 [3].

Elements	Vandetanib
C	55.59%
Br	16.81%
N	11.79%
O	6.73%
H	5.09%
F	4.00%

1.4 Appearance

Vandetanib appearance is a white to off-white powder [4].

2. Physical characteristics

2.1 Color/form

The color of Vandetanib powder is white to off-white. [4].

2.2 Flash point

279.3 °C [5].

2.3 Melting point

242 °C [6].

2.4 Dissociation constant

The dissociation constant of Vandetanib was evaluated via the OECD 112 method and it was found to be pKa$=5.14$ (NH Group). On the other hand, the dissociation constant of Vandetanib was predicted using ACD Software for N with an adjoining methyl group and it was found to be pKa (strongest basic)$=9.13$. [7]

2.5 Solubility and partition coefficient

In terms of solubility, Vandetanib has a pH-dependent behavior, with higher solubility at lower pH. Vandetanib has a solubility in water of 0.008 mg/mL at 25 °C (77 °F), which is nearly insoluble in water [8]. Using the OECD 107 procedure, the partition coefficient (octanol/water) of Vandetanib is optioned as following: log $D_{ow} = -0.684$ at pH 3, log $P = 2.21$ at pH 7, and log $P > 3.90$ at pH 11. [9]

2.6 Particle morphology

Vandetanib powder was structurally studied by high-resolution X-ray diffraction (XRD) utilizing a Rigaku Ultima IV diffractometer with a scintillation detector in reflection mode, Cu K{acute over (α)} 19.400 radiation (1.5406A°), Scanning range: 3–60° 2θ, Step size: 0.02° 2θ and Time per step: 1 s, monochromatized with a graphite crystal. The pattern was collected at 40 kV of tube voltage and 40 mA of tube current in step scan mode. The peaks (reflections) of vandetanib powder are obtained at Table 3 and shown in Fig. 2.

2.7 Thermogravimetric analysis (TGA)

Thermal Gravimetric Analysis (TGA) of vandetanib was obtained using a PerkinElmer pyris 1 apparatus. An Aluminum pan containing 2.152 mg of sample was perforated prior to scanning, and a temperature profile of 50–550 °C was run at a rate of 10 °C/min under a nitrogen purge (50 mL/min). Figure 3 shows degradation profiles that clearly demonstrate the mass loss over the course of the degradation period. But it's clear that the degradation process only happens at one stage. And the compound was decompose at 253.83 °C. as shown in Fig. 3.

Table 3 The X-ray powder diffraction pattern of vandetanib.

Peak no.	2theta	Flex width	d-Value	Intensity	I/Io
1.	8.300	0.353	10.6440	618	20
2.	12.600	0.235	7.0195	289	10
3.	15.100	0.235	5.8625	3118	100
4.	16.300	0.118	5.4335	530	17
5.	18.200	0.235	4.8703	1163	38
6.	19.000	0.235	4.6670	1070	35
7.	20.800	0.235	4.2670	1378	45
8.	21.200	0.118	4.1874	389	13
9.	21.500	0.235	4.1297	2849	92
10.	22.200	0.235	4.0010	584	19
11.	23.400	0.235	3.7985	1392	45
12.	23.900	0.235	3.7201	950	31
13.	24.300	0.235	3.6598	296	10
14.	25.100	0.235	3.5449	443	15
15.	25.400	0.235	3.5037	320	11
16.	25.700	0.118	3.4635	128	5
17.	26.900	0.235	3.3117	320	11
18.	27.400	0.235	3.2524	213	7
19.	27.700	0.118	3.2178	123	4
20.	29.600	0.353	3.0154	633	21
21.	30.100	0.118	2.9665	94	3
22.	30.400	0.353	2.9379	285	10
23.	31.900	0.235	2.8031	134	5
24.	32.900	0.235	2.7201	310	10
25.	33.400	0.353	2.6805	83	3
26.	35.900	0.353	2.4994	90	3
27.	36.600	0.235	2.4532	121	4
28.	36.900	0.353	2.4339	173	6

Continued

Table 3 The X-ray powder diffraction pattern of vandetanib.—cont'd

Peak no.	2theta	Flex width	d-Value	Intensity	I/Io
29.	38.500	0.118	2.3364	82	3
30.	40.000	0.353	2.2522	116	4
31.	41.900	0.235	2.1543	143	5
32.	43.400	0.353	2.0833	110	4
33.	45.200	0.353	2.0044	113	4

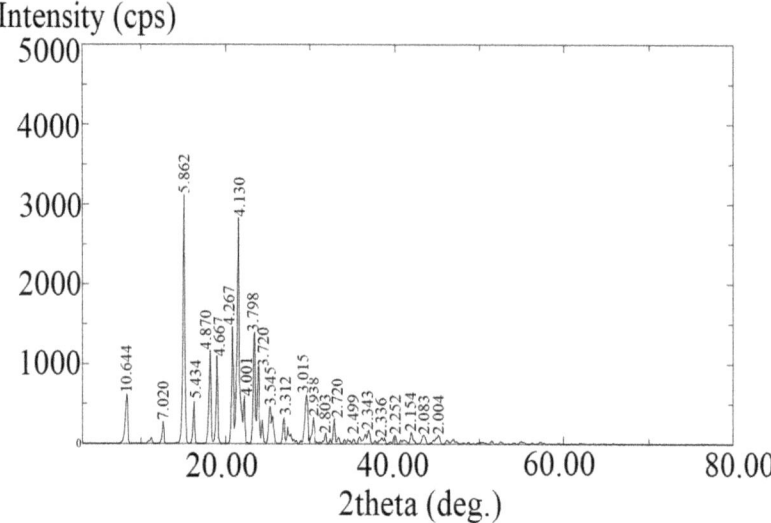

Fig. 2 The X-ray powder diffraction pattern of vandetanib.

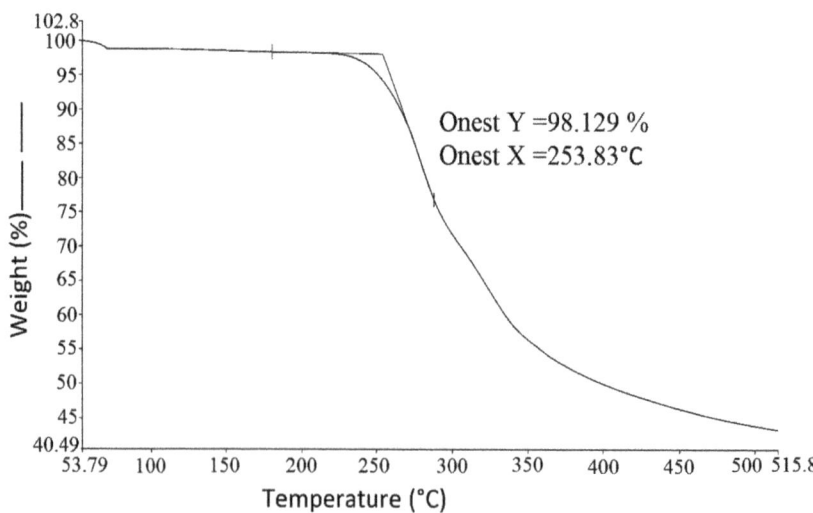

Fig. 3 Thermogravimetric analysis (TGA) of vandetanib.

3. Methods of preparation

Hennequin et al. [10] developed a method to prepare vandetanib starting from reacting aminobenzamide (1) with Gold's reagent in the presence of dioxane and reflux to achieve (2). Then (2) **they** react with thionyl chloride in DMF and reflux to give 4-chloroquinazoline (3). Nucleophilic substitution of the chlorine atom of (3) with (4) yielded 4-anilinoquinazoline (5), followed by cleavage of the C-7 benzyl protecting group with trifluoroacetic (TFA) under reflux, leading to unprotected C-7 hydroxy-4-anilinoquinazoline (6), which is then subjected to direct alkylation of the phenol moiety with (7) in DMF, affording (8). Vandetanib is finally obtained by dissolving (8) in HCHO and HCOOH at 95 °C (Scheme 1).

An alternate technique of synthesizing vandetanib through the Dimroth rearrangement was documented [11], which used a microwave to speed up the process. A 7% yield of Vandetanib was synthesized in over 9 stages of the novel approach, compared to 4%–20% yield over 12–14 steps in the previously published methods. Scheme 2 illustrates this. The synthesis began with the cheap building block 4-hydroxy-3-methoxybenzonitrile 1, which was quantitatively alkylated with benzyl bromide without the need for substantial purification. Under moderate circumstances, nitration of 2 yielded 3 in a 93% yield as a yellow precipitate that required no additional purification.

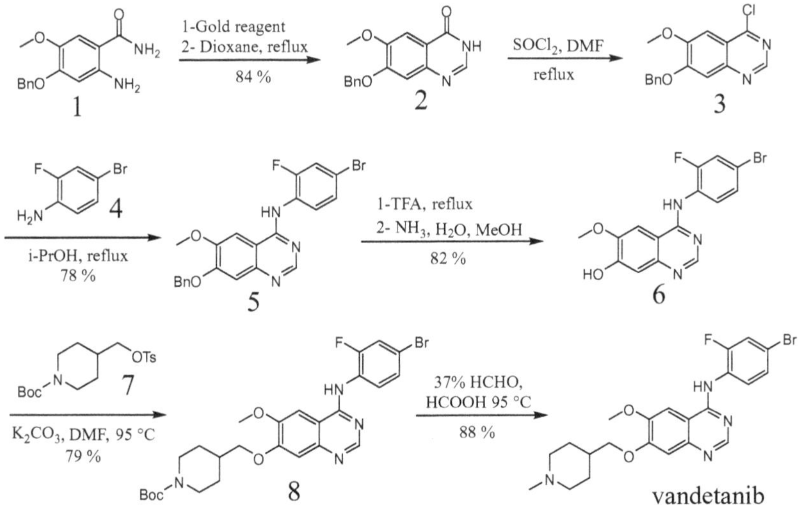

Scheme 1 Synthesis of vandetanib.

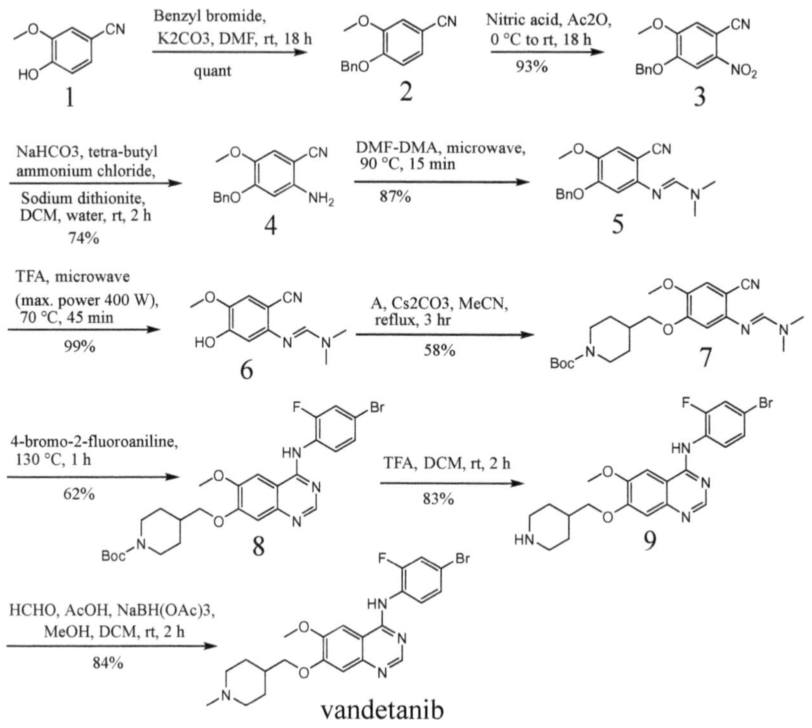

Scheme 2 Synthesis of vandetanib through the Dimroth rearrangement.

The TFA-mediated debenzylation of **5** was chosen because of the formamidine's possible sensitivity to hydrogenation–induced reduction and yielded **6** in a virtually quantifiable yield. In 58.6% yield, alkylation of 6 with tert-butyl-4-(tosyloxy) methyl) piperidine-1–carboxylate yielded 7. The crucial Dimroth rearrangement step was next examined, beginning with **7** in the presence of 4-bromo-3-fluoroaniline (1 equivalent) and acetic acid, with microwave heating at 130 °C for 45 min. The duration of the microwave irradiation was unimportant for the synthesis of **8**, which was temperature sensitive and needed 130 °C for full conversion, underlining the possibility of inactivated reagents. After chromatographic purification, quinoline **8** was recovered in a 62% yield. The BOC deprotection of piperidine **8** was the final step in the synthesis of vandetanib. However, deprotection using the approaches described in the literature resulted in poor yields during aqueous work-up. However, by following the deprotection process described in the literature and avoiding water work-up, straight silica chromatography purification yielded a **9** in 83.1% yield. The literature technique says that reductive amination with formaldehyde gave vandetanib at an 84.4% yield, for a total yield of 7% over nine stages.

4. Spectroscopy identification

4.1 Ultraviolet spectroscopy

A Shimadzu UV-spectrophotometer (model no. UV-1800) was used to record an ultraviolet absorption spectrum of 20 μg/mL vandetanib in methanol. The solution's absorption spectra were measured in a 1 cm quartz cell within the 200–400 nm range. Fig. 4 shows that vandetanib has two places in the spectrum where it has maximum absorption: at 249.51 and 330.31 nm.

4.2 Spectrofluorometry

The fluorescence spectrum emission of vandetanib in methanol shown in Fig. 5 was recorded using a Jasco FP-8200 Spectrofluorometry (Jasco Corporation, Japan) equipped with a 150 W xenon lamp and 1 cm quartz cells. The slit widths for both the excitation and emission monochromators were set at 5.0 nm. The fluorescence spectrum of vandetanib is shown in

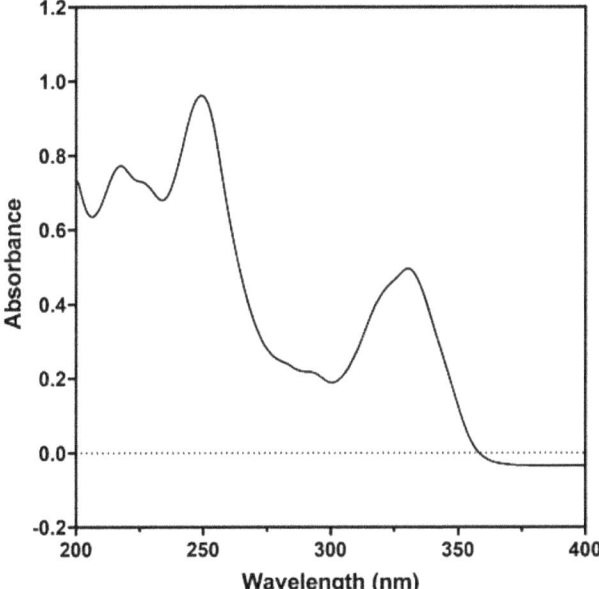

Fig. 4 The UV- absorption spectrum of vandetanib.

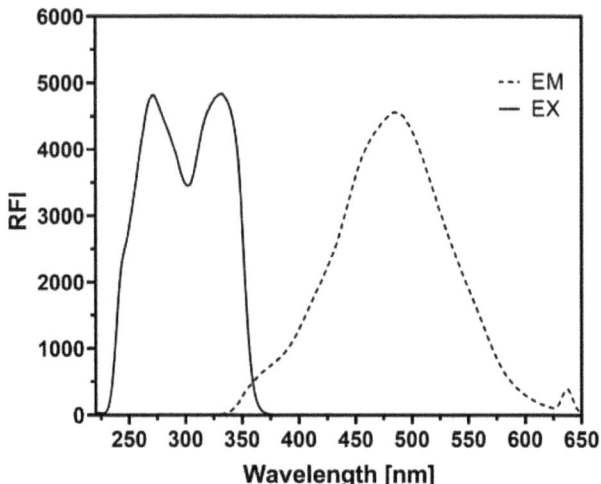

Fig. 5 The fluorescence spectrum of vandetanib.

Fig. 5, which exhibited maximum excitation at 270.21 and 329.65 and maximum emission at 486.17 nm.

4.3 Infrared spectroscopy

The infrared absorption spectrum of vandetanib was recorded as a KBr disk using the Perkin Elmer FT-IR Spectrum BX apparatus. Figure 6 shows the FT-IR spectrum from 400 to 4000 cm^{-1} of vandetanib. The characteristic absorption bands of vandetanib are shown in Table 4.

4.4 Mass spectrometry

The mass spectrum of vandetanib ($C_{22}H_{24}BrFN_4O_2$, 475.4) was obtained using an Agilent 6320 Ion Trap Mass Spectrometer (Agilent Technologies, USA) equipped with an electrospray ionization interface (ESI). Solubility of Vandetanib was shown in DMSO (27.5 mg/mL; needed ultrasonic and warming), so a stock solution of vandetanib was prepared in DMSO to 1 mg/mL and diluted with 50:50 acetonitrile and water (HPLC grade) to 1 μg/mL. The diluted solution was injected into the ion trap mass spectrometer by direct infusion using a connector instead of the HPLC column. The flow rate of the mobile phase was 0.4 mL/min and the total run time was 5 min. The MS parameters were optimized, and the scan was in ultra-scan mode. The ESI was operated in positive mode. The source temperature was set to 350 °C with a nebulizer gas pressure of 55.00 psi and a dry gas

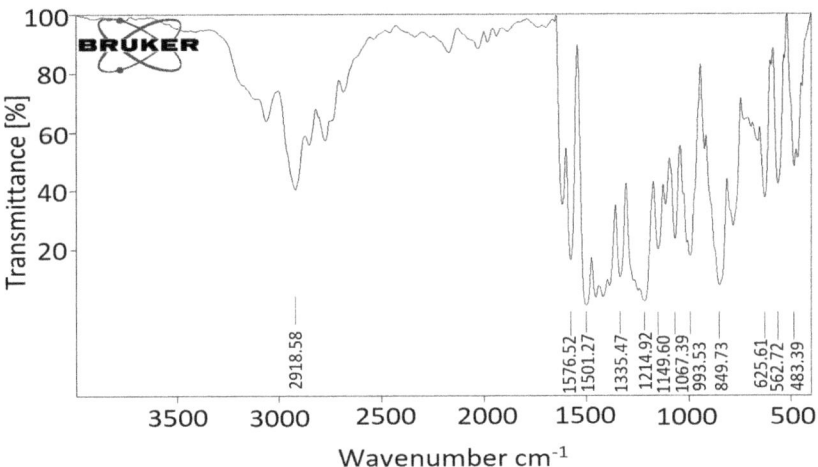

Fig. 6 IR spectrum of vandetanib.

Table 4 Interpretation of vandetanib FTIR spectra.

No.	Functional Group	Standard Value	Obtained Value
1	C-O group	$1050–1150 \, cm^{-1}$	1067.39
2	C-N group	$1000–1350 \, cm^{-1}$	1335.47
3	C=C group	$1470–1590 \, cm^{-1}$	1576.52
4	C-H group	$2850–3000 \, cm^{-1}$	2918.58
5	N-H group	$3060–3500 \, cm^{-1}$	~3100

flow rate of 12.00 L min. Figure 7A represents the full mass scan at a m/z range from 50 to 750, showing the molecular ion peak at $m/z = 475.2 \, [M+H]^+$. The Vandetanib chemical structure contains bromine atoms that result in generating an isotopic pattern at m/z 475.2 and m/z 477.2 (Fig. 7B). Fragmentation of the molecular ion peak at m/z 475.2 generates product ions at m/z 442.2 and 400.0 (Fig. 7C). The sequential breakage of the methyl N-methyl piperidine moiety is represented by fragments at m/z 442 and 400 (Scheme 3).

4.5 NMR spectroscopy

4.5.1 ^1H NMR spectroscopy

A Bruker NMR spectrometer running at 500 MHz was used to record the ^1H NMR spectrum of vandetanib in CDCl$_3$. Chemical shifts are represented

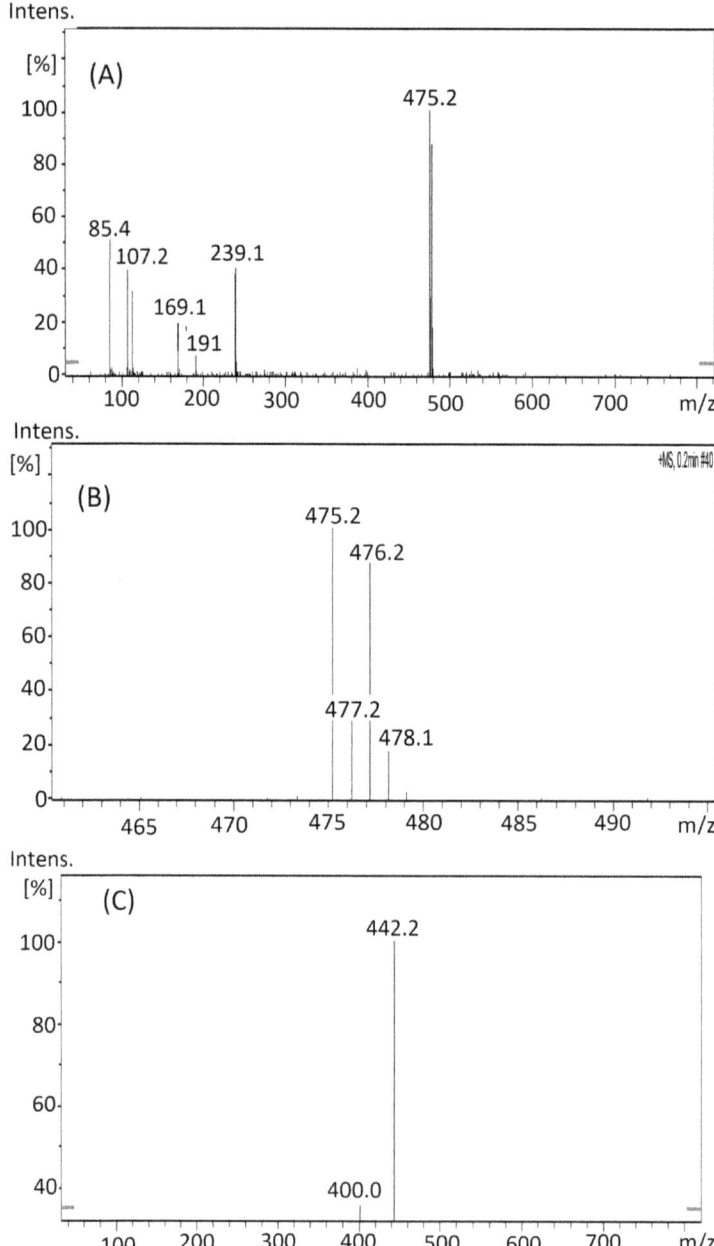

Fig. 7 (A) Full mass scan of vandetanib. (B) Isotopic pattern of vandetanib's molecular ion peak (C) fragmentation mass spectrum of Vandetanib molecular ion peak at *m/z* 475.

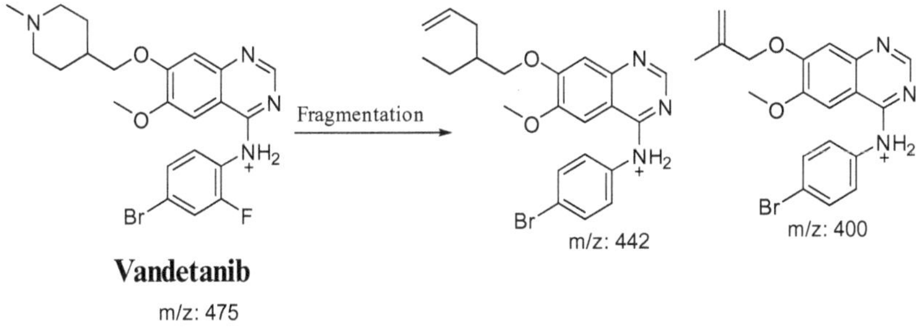

Vandetanib
m/z: 475

Scheme 3 Proposed fragmentation pattern of vandetanib.

Table 5 ^{1}H NMR of vandetanib (CDCl$_3$).

	Location (δ) ppm	Shape	Integration	Correspondences
1	1.35	q	2H	–CH$_2$ in piperidine
2	1.76	m	3H	–CH–CH$_2$ in piperidine
3	1.86	t	2H	–N–CH$_2$ in piperidine
4	2.16	s	3H	–N–CH$_3$
5	2.79	t	2H	–N–CH$_2$ in piperidine
6	3.95	s	3H	–O–CH$_3$
7	4.01	d	2H	–O–CH$_2$- piperidine
8	7.18	d	1H	Ar CH in benzene
9	7.46	d	1H	Ar CH in benzene
10	7.53	s	1H	Ar CH in benzene
11	7.66	s	1H	Ar CH in quinazoline
12	7.80	s	1H	Ar CH in quinazoline
13	8.36	s	1H	–NH–
14	9.56	s	1H	–CH in pyrimidine

as δ-values (ppm) relative to the internal standard (TMS). The coupling constants (J) are measured in hertz (Hz) (Table 5 and Fig. 8).

4.5.2 ^{13}C NMR spectroscopy

^{13}C NMR Spectrum of vandetanib was recorded in CDCl$_3$ on a Brucker NMR spectrometer operating at 125 MHz. Chemical shifts are expressed in δ-values (ppm) relative to TMS as an internal standard (Table 6 and Fig. 9).

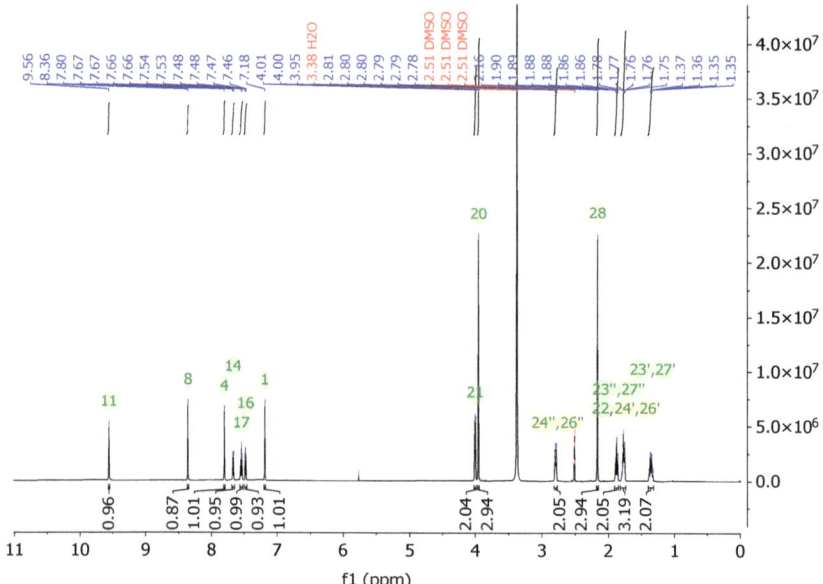

Fig. 8 ^{1}H NMR spectrum of vandetanib.

Table 6 ^{13}C NMR of vandetanib (DMSO-d_6).

Signal	Location (δ) ppm	Integration	Correspondences
23, 27	28.96	2C	$-(CH_2)2$ in piperidine
22	35.08	1C	$-CH-$ in piperidine
28	46.69	1C	$-N-CH_3$
24, 26	55.34	2C	$-N-(CH_2)2$ in piperidine
20	56.63	1C	$-O-CH_3$
21	73.25	1C	$-O-CH_2-$ piperidine
4	102.41	1C	$-CH-$ in quinazoline
1	108.09	1C	$-CH-$ in quinazoline
3	109.02	1C	$-C-$ in quinazoline
15, 14	117.98, 119.73	2C	$-CH-C-Br$ in benzene
17, 16, 12	126.85, 127.96, 130.01	3C	$-N-C-CH-CH$ in benzene
2	147.41	1C	$-C-N-$ in quinazoline
6	149.57	1C	$-C-OCH_3$ in quinazoline
8	153.38	1C	$-N-CH=N-$ in quinazoline
5	154.26	1C	$-C-O-$ in quinazoline
13	156.41	1C	$-C-F$ in benzene
10	157.35	1C	$N=C-NH-$ in quinazoline

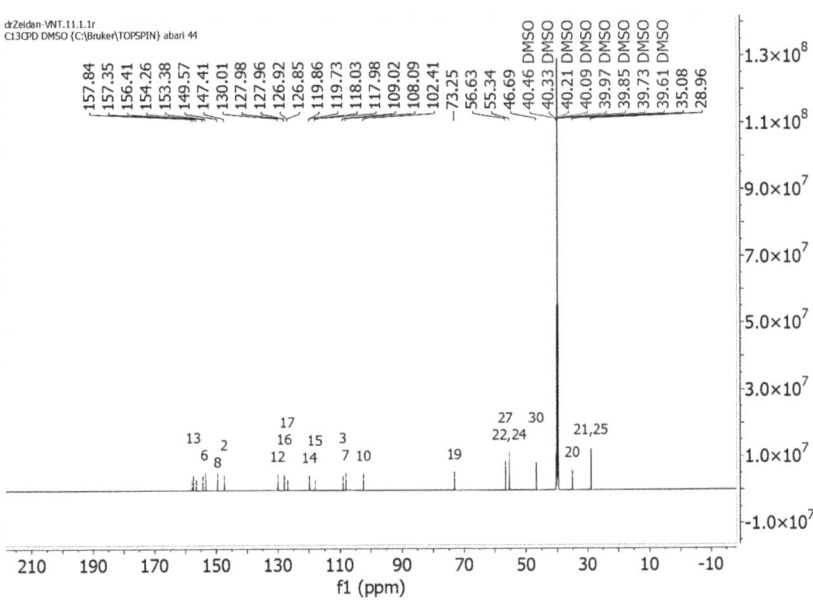

Fig. 9 ^{13}C NMR spectra of vandetanib.

5. Method of analysis

5.1 Spectroscopic methods

5.1.1 Ultraviolet-visible spectroscopy

Khandare et al. [12] developed and validated a simple, precise and economical UV-spectrophotometric method for the determination of vandetanib from bulk. Two methods were developed. The first approach used was area under curve (AUC), which integrated the wavelength range of 323.59–333.36 nm. The second method (B) used first order derivative spectrometric analysis. This method measured absorbance at $\lambda_{min} = 311.27$ nm, $\lambda_{max} = 340.54$ nm, and zero cross $= 328.37$ nm. Calibration curves for the technique were plotted using instrumental response at selected wavelengths and analyte concentrations in the solution. The linearity of the concentration ranges for both methods from 5 to 30 µg/mL at $\lambda_{max} = 328.44$ nm and the correlation coefficient of 0.9966 imply that Beer's- Lambert's Law applies.

The drug recovered well at each of the 80%, 100%, and 120% levels, with recoveries ranging from 97.00% to 99.00% for both procedures, implying that the approach was accurate. The accuracy, precision, limit of detection (LOD), and limit of quantitation (LOQ) of the analytical technique were determined. The methods were validated according to the International Conference Harmonization (ICH) guidelines. Every validation parameter was within the permitted range. The proposed approach was used effectively to measure the quantity of vandetanib in pharmaceutical formulations.

Abdelhameed et al. [13] created and validated six successive spectrophotometric-based univariate approaches for the simultaneous quantification of three new anticancer medicines in a mixture: vandetanib, dasatinib, and sorafenib. These procedures are unique, straightforward, precise, and accurate. These analytical methods were developed through a series of steps that included zero crossing, ratio-based, and/or derivative spectra. The names of these methods are the ratio difference spectrophotometric method, the constant center method, the successive derivative ratio method, the isoabsorptive method, the mean centering of the ratio spectra method, and the derivative ratio spectrum-zero crossing method. The linearity of the calibration curve varied from 2 to 9, 2 to 9, and 3 to 9 µg/mL for vandetanib, dasatinib, and sorafenib, respectively, with a correlation coefficient (r) of ≥ 0.9997. These well-established procedures were used to quantify the three medicines in various biological fluids (spiked human plasma and urine) and pharmaceutical preparations. These techniques are faster than other chromatographic methods such as HPLC-UV and LC-MS, and the findings did not differ significantly from the other described methods, showing no significant difference in accuracy and precision.

Also Abdelhameed et al. [14] developed rapid, accurate, and validated new univariate and multivariate chemometric spectrophotometric techniques for a simultaneous determining mixture of vandetanib, dasatinib, and sorafenib in pure form, tablets, and spiked human plasma and urine. Also, it can be applied to the dissolution profiles for these drugs. These methods comprised the double divisor ratio spectra derivative univariate method as well as the chemometric multivariate method, which featured partial least squares (PLS) and principal component regression (PCR). The Double divisor ratio spectra derivative absorption minimum at 358.4 nm was used to quantify vandetanib, absorption maximum at 300.3 nm was used to quantify dasatinib, and absorption maximum at 259.8 nm was used to quantify sorafenib. This approach demonstrated linearity in the concentration ranges of 2–9 µg/mL for vandetanib and dasatinib, and 3–9 µg/mL for sorafenib with a correlation value (r^2) of ≥ 0.9998. Chemometric PLS and PCR models were found to be linear in the same ranges with a correlation value (r^2) of ≥ 0.9997. No significant difference was found between the proposed and reported method.

5.1.2 Spectrofluorimetric method

For the detection of vandetanib in tablets and biological fluids (spiked human plasma and urine), Darwish et al. [15] developed and validated a spectrofluorimetric technique. The suggested approach is based on measuring vandetanib's intrinsic fluorescence intensity in acetonitrile at 480 nm upon excitation at 330 nm. The impact of pH, diluting solvent, and time on vandetanib fluorescence intensity was evaluated and adjusted using a one component at a time method. By graphing vandetanib fluorescence intensity at 480 nm versus vandetanib concentrations in ng/mL, a calibration curve was generated. The approach was validated in accordance with the International Conference on Harmonization (ICH) standards for the validation of analytical processes. The linearity range of the technique was $20,600 \, \text{ng} \, \text{mL}^{-1}$, with quantification (LOQ) and detection limits (LOD) of 30.45 and 10.05 $\text{ng} \, \text{mL}^{-1}$, respectively. The proposed approach was successfully applied to the quantification of vandetanib in pure powder form ($100.90 \pm 0.91\%$), laboratory prepared tablets ($97.86 \pm 1.42\%$), spiked human plasma ($97.97 \pm 2.36\%$) and urine ($97.59 \pm 0.87\%$). When the suggested approach was compared to that of liquid chromatography-tandem mass spectrometry, there was no significant difference ($P < 0.05$) in terms of accuracy and precision. As a conclusion, the suggested approach can be used to analyze vandetanib in biological samples and in dosage form.

An RSM-assisted micellar improved synchronous spectrofluorimetric technique for the measurement of vandetanib (VDB) in tablets, plasma, and urine was developed and validated by Darwish et al. [16]. The method involved enhancing the fluorescence behavior of VDB in polyoxyethylene hydrogenated castor oil 40 (HCO40) micellar medium and detecting fluorescence utilizing synchronous scan methodology ($\Delta\lambda = 50\,nm$). RSM used the Box–Behnken design to optimize key parameters controlling VDB fluorescence. The type and volume of surfactant, as well as the buffer medium's pH, were the determining factors. The fluorescence–concentration plot was linear over the 40–600 ng mL^{-1} range under ideal circumstances, with detection and quantification limits of 5.22 and 15.82 ng mL^{-1}, respectively. The proposed method was used to analyze laboratory-prepared tablets, spiked human plasma, and urine samples with great success. It was discovered that the results from this method and those from a pre-validated liquid chromatography-tandem mass spectrometric reference method agreed well statistically.

5.2 Chromatographic methods

5.2.1 High-performance liquid chromatography methods

Lin et al. [17] developed and validated a simple and sensitive high-performance liquid chromatography (HPLC) method with ultraviolet detection for quantification of vandetanib in rat plasma. Vandetanib and the internal standard (I.S.) trazodone hydrochloride was separated using a C18 Atlantis column and a gradient mobile phase of acetonitrile/0.5% triethylamine with pH 3.0 and a flow rate of 1.0 mL/min. The detection wavelength was 341 nm. The calibration curve range was 80–4000 ng/mL with an R2 of 0.9998 was achieved. The intra- and inter-assay accuracy ranged from 98.80% to 103.08% and 95.32% to 98.40%, respectively, with high precision (R.S.D.% 5%). The mean absolute recovery was 96.65%. This approach is adequate for vandetanib pharmacokinetic research in small animals and may be used in human pharmacokinetic studies.

For the simultaneous analysis of four tyrosine kinase inhibitors in various human plasma samples, a simple approach was invented by Xiang et al. [18] using high-performance liquid chromatography (HPLC) with diode array detection (DAD) as the detector and the alternating trilinear decomposition method. The analytes were separated by chromatography on a reversed-phase column (C18: 200 mm × 4.6 mm, 5.0 μm) with methanol (65%, v/v, A) and 0.1% formic acid aqueous solution (35%, v/v, B). Each run of analysis took 5 min, and analytes could be fully eluted in 2.8–3.8 min. Vandetanib, pazopanib, afatinib, and dasatinib calibration concentration ranges were designed as 0.50–6.10, 0.50–6.10, 0.70–7.00, and 0.70–7.00 μg/mL respectively. Intra- and inter-day RSDs varied from

0.1 to 8.9%. With the use of the "second-order advantage" of three-way (second-order) calibration procedures, quantitative information may be derived from the unsegregated interferences of distinct human plasma samples. All of the results showed that the proposed method for direct quantitative analysis of the four tyrosine kinase inhibitors in different complex systems had good characteristics of rapidity, sensitivity, and efficiency, and it is assumed to be an appealing choice in the fast analysis of clinical samples.

Khandare et al. [19] developed and validated a method for the determination of vandetanib in bulk dosage form. The method was fast, simple, sensitive, precise, and reproducible using reversed phase high performance liquid chromatography (RP-HPLC). The separation was done using a reversed phase C18 column (Inerstil ODS-3 V 5 um 250 × 4.6 mm) under ambient temperature. The mobile phase consists of methanol (100 v/v) at a rate of 1 mL/min. The wavelength of the UV detector was 328 nm. The method was validated according to many parameters like precision, limit of quantitation (LOQ), linearity, and robustness. The linearity of the method was found to be linear over the range of 50–100 µg/mL with r2 = 0.9996. The retention time for bulk vandetanib was found to be 5.496 ± 25 min. The method's LOQ was 6.8339 g/mL and its LOD was 2.7036 g/mL. Thus, the developed method was found to be robust and rugged, which can be used for the regular analysis of vandetanib in the bulk and pharmaceutical dosage form.

Salode et al. [20] developed and validated a reverse phase high performance liquid chromatography method (RP-HPLC) for the determination of vandetanib. The separation was done using a Nucleosil 100–5 C18 (250 × 4.6 mm, 5 µm) column and methanol: ammonium acetate buffer as mobile phase (90:10 v/v) with a flow rate of 1 mL/min. The detection wavelength was at 249 nm and the retention time (R_t) of vandetanib was found to be 3.717 ± 0.034 min. The method was validated according to ICH guidelines. It gives a good linear relationship. The correlation coefficient was found to be 0.992 in the concentration range of 1–10 µg/mL. The % assay was found to be 100.144 ± 1.032. Furthermore, vandetanib was subjected to various stress conditions like acidic, alkali, oxidative, thermal, and photolytic degradation. The degradation pathways for major degradants were identified and characterized by the Liquid Chromatography Tandem Mass Spectrophotometric method (LC-ESI-MS) equipped with Q-TOF. The drug and its degradation products were quantified using MS detection in positive ion mode. In acquisition mode, the minimum range was set to

60 and the maximum range was set to 1000, with a scan rate of 2 spectra/sec. This method shows stability and can be used for routine drug analysis, both in bulk form and in the form of pills.

Murali, and Venkateswara Rao [21] established and validated an isocratic RP-HPLC technique for the fast assessment of Vandetanib, an anticancer medication, in both bulk and pharmaceutical dosage forms. At room temperature, a symmetric C18 column was eluted at a flow rate of 1 mL/min. The mobile phase was composed of acetonitrile, water, and orthophosphoric acid in the proportions of 90:08:02 (v/v/v). In the concentration range of 50 to 200 ppm, linearity was observed. Vandetanib had a retention time of 3.326 min. The approach was validated in accordance with ICH recommendations. The proposed approach is applicable to the determination of Vandetanib in pharmaceutical dosage forms. In addition, the detection of Vandetanib was independently validated using LC–MS using the ESI method, allowing for further research of this drug using the LC–MS approach as well.

5.2.2 Liquid chromatography mass spectrometry methods

A liquid chromatography tandem mass spectrometry technique was used by Zirrolli et al. [22] to determine vandetanib levels in mouse plasma and tissues. A fast and sensitive method was developed using the HPLC system coupled with a triple quadrupole mass spectrometer. The mass spectra were obtained using positive ion electrospray ionization (ESI). After adding the internal standard (trazodone), plasma (0.05 mL) and tissue homogenates (0.1 mL of 10 mg/mL) were extracted under alkaline conditions with ethyl acetate: pentane (1:1, v/v). Separation was accomplished on a C18 column (50 mm × 2 mm, 5 μm), with quantification by internal standard reference and repeated reaction monitoring of the ion transitions m/z 475 → 112 (vandetanib) and m/z 372 → 176 (trazodone). The calibration curve was linear throughout a 20–20,000 ng/mL range in plasma and a 10–320 ng/mg range in tissue homogenates. The mean recovery rates from plasma and tissue homogenates were 88% and 90%, respectively. The accuracy in plasma was 88% with good precision (R.S.D. < 10%) at the lowest limit of quantification (20 ng/mL with a 50 L plasma sample). In a pharmacokinetic study on mice, this method was used to find doses of vandetanib that would be similar to the therapeutic values found in people.

Bai et al. [23] developed and validated a highly sensitive and precise method for the determination of vandetanib (ZD6474) in human plasma and cerebrospinal fluid (CSF) by liquid chromatography electrospray ionization tandem mass spectrometry (LC-ESI-MS/MS) using [(13)C, d(3)]-ZD6474 as an internal standard (ISTD). Samples were prepared by liquid-liquid extraction with tert-butyl methyl ether containing 0.1% or 0.5% ammonium hydroxide. ISTD were separated using a Kinetex C18 column (2.6 μm, 50 mm × 2.1 mm) at ambient temperature with an isocratic mobile phase (acetonitrile/10 mM ammonium formate = 50/50, v/v, at pH 5.0) with a flow rate of 0.11 mL/min. The retention time of both compounds was 1.60 min in a runtime of 3 min. Detection was achieved by an API-3200 LC-MS/MS system consisting of a triple quadrupole mass spectrometer, using positive multiple-reaction monitoring (MRM) mode, monitoring m/z 475.1/112.1 and m/z 479.1/116.2 for vandetanib and ISTD, respectively. The linearity of the method at range of 0.25–50 ng/mL was good with (R2 $\geq$ 0.990) for the CSF curve and from 1.0 to 3000 ng/mL with (R^2 $\geq$ 0.992) for the plasma curve. The mean recovery for vandetanib was 80%. Inter-day and intra-day precision were $\leq$8.8% and $\leq$5.9% for CSF and plasma, respectively. Inter-day and Intra-day accuracies ranged from 95.0% to 98.5% for CSF, and from 104.0% to 108.5% for plasma. When tested in plasma from six different sources, Vandetanib had no matrix impact. This method was used to look at pharmacokinetic samples taken from kids with brain tumors who were taking vandetanib by mouth.

Andriamanana et al. [24] developed and validated a method for the simultaneous determination of vandetanib and other anticancer agents in human plasma by liquid chromatography-tandem mass spectrometry (LC-MS/MS), performed by electrospray ionization in positive mode using a triple quadrupole mass spectrometry analysis. Separation was achieved by the Hypersil Gold ($^®$) PFP column using a gradient elution of 10 mM ammonium formate containing 0.1% formic acid (A) and acetonitrile containing 0.1% formic acid (B) at a flow rate of 0.3 mL/min. After addition of the internal standard and protein precipitation, the supernatant is diluted 2-fold in a mixture of A and B (50/50, v/v). The standard curves range from 50 to 3500 ng/mL. The method showed good results in terms of sensitivity, specificity, precision (intra- and inter-day RSD from 3.7% to 13.8%), accuracy (from 86.8% to 113.5%), and stability of the analytes under various conditions.

Amer et al. [25] developed and validated a fast, specific, and sensitive liquid chromatography tandem mass spectrometry method for the quantification of vandetanib in human plasma and rat liver microsome matrices. This method was used for the metabolic stability investigation of vandetanib. The separation was performed using a C18 column and an isocratic mobile phase composed of 10 mM ammonium formate (pH 4.1) and acetonitrile in a ratio of 1:1. The flow rate was set at 0.25 mL/min and the total run time was 4 min with an injection volume of 5 μL. ESI was the ion source which analyzed it using the multiple reaction monitoring mode (basis for quantification) in the Agilent 6410 QqQ analyzer. The linearity ranged from 5 to 500 ng/mL ($r^2 \geq 0.9996$) in human plasma and rat liver microsomes. LOQ and LOD were 2.48 and 7.52 ng/mL, and 2.14 and 6.49 in human plasma and rat liver microsome matrices, respectively. The intra-day and inter-day precision and accuracy were 0.66–2.66% and 95.05–99.17% in the human plasma matrix, while in the RLMs matrix, they ranged from 0.97 to 3.08% and 95.8 to 100.09%, respectively.

Denys et al. [26] reported a method to evaluate the pharmacokinetics, safety, and toxicity following intra-arterial hepatic artery administration of vandetanib-eluting radiopaque beads in healthy swine. The method used solid phase extraction followed by liquid chromatography coupled with tandem mass spectrometry (LCMS/MS) to determine vandetanib and its metabolite in swine plasma and liver using turbo spray in positive ion with multiple reaction monitoring mode. Separation was achieved using a 5 mL injection into a Thermo Accucore HILIC (50 × 3 mm, 2.6 μm) analytical column run at 40 °C, using formic acid (0.2%) in acetonitrile: water: ammonium formate (1 M) mobile phases at 0.6 mL/min. Assay Standards: Vandetanib together with various metabolites and labeled analytical/internal standards, were used as reference standards for the chromatography.

Attwa et al. [27] reported a method for the characterization and identification of in vitro, in vivo and reactive metabolites of vandetanib by liquid chromatography electrospray ionization tandem mass spectrometry (LC–ESI–MS/MS). The separation was performed using the Agilent eclipse plus C_{18} Column (1.8 μm, 50 mm × 2.1 mm) for in vitro and (3.5 μm, 150 mm × 2.1 mm) for in vivo. The gradient mobile phase is composed of 10 mM ammonium formate in H_2O (pH: 4.1 using formic acid) and acetonitrile (0.1% formic acid). The flow rate was set at 0.2 mL/min and the total run time was 90 min. Positive electrospray ionization (ESI) was the ionization source and the analysis was performed using product ions (PI) and full mass scan in the Agilent 6410 QqQ analyzer.

6. Stability

According to EU/ICH stability guidelines, vandetanib was found to be stable in thermal and hydrolytic conditions, but there was a little degradation within photolytic conditions. In solution, vandetanib is degraded under acidic, oxidative, and light stress conditions, but it is stable under basic conditions [4].

The study by Dall'Acqua et al. [28] was the first reported of photodegradation of vandetanib. When aqueous solutions of the anticancer medication vandetanib are exposed to UV-A light, they undergo photochemical destruction. HPLC–MS analysis found two major photodegradation products, and their structures were explored using LC–MS and NMR spectra following their separation by HPLC. The photoproducts of simple debromination (N-(2-fluorophenyl)- 6-methoxy-7-((1-methylpiperidin-4 yl)methoxy)quinazolin-4-amine, FP3) or bromide loss followed by solvent addition (N-(4-hydroxy-2-fluorophenyl)-6-methoxy-7-((1-methylpiperidin-4 yl)methoxy) quinazolin-4- amine, FP2).

7. Pharmacology

7.1 Dosage form
Oral tablet 100 and 300 mg [8].

7.2 Mechanism of action

Vandetanib is a selective inhibitor of multiple tyrosine kinases, including vascular endothelial growth factor receptor-2 (VEGFR-2), epidermal growth factor receptor (EGFR), and rearranged during transfection (RET) receptor tyrosine kinases. In vivo, vandetanib decreases tumor cell-induced angiogenesis, tumor vessel permeability, and inhibits tumor growth and metastases [29].

7.3 Pharmacokinetics

7.3.1 Absorption
According to Frampton [30] According to Frampton [28] vandetanib is taken orally and slowly absorbed, so maximum plasma concentrations ($C_{max} = 857$ ng/mL) of the drug are reached after dose within a median time of 6 h. Vandetanib accumulates approximately 8-fold on multiple dosing, with steady state achieved after $\approx$2 or 3 months. C_{max} and AUC0–∞ (area under the plasma concentration-time curve from time zero to infinity) did not significantly affect after a high-fat meal with single dose of vandetanib (300 mg); that's mean the drug can be taken with or without food.

7.3.2 Distribution

As reported by Frampton [30] vandetanib has high affinity binding to plasma proteins, approximately 92–94% with a 7450 L volume of distribution. Vandetanib can bind to human serum albumin and α1-acid-glycoprotein.

7.3.3 Metabolism

Frampton [30] found in his research that vandetanib is metabolized to N-desmethyl vandetanib by the cytochrome P450 (CYP) 3A4 isoenzyme and to vandetanib N-oxide by flavine-containing mono-oxygenases in the kidney (FMO1) and liver (FMO3). Both of these metabolites, together with the unchanged parent drug, were detected in the plasma, urine, and feces of healthy volunteers after oral administration of radio-labeled vandetanib (single 800 mg dose). A third metabolite (a glucuronide conjugate of vandetanib) was detected in the urine and feces only. Exposure to N-desmethylvandetanib and vandetanib N-oxide relative to the parent drug was 7–17.1% and 1.4–2.2%, respectively. In terms of in vitro inhibition of VEGFR-2 and EGFR, the potency of N-desmethylvandetanib is similar to that of vandetanib, whereas vandetanib N-oxide is >50-fold less active than the parent drug.

According to BC Cancer Drug Manual [29], N-desmethylvandetanib and vandetanib N-oxide are active metabolites, and the glucuronide conjugate of vandetanib is an inactive metabolite.

7.3.4 Elimination

According to Frampton [30] According to Frampton [28] biliary and urinary excretion are both important routes of elimination for vandetanib. Approximately 69% of the orally administered radiolabeled dose (800 mg) was recovered in the feces (44%) and urine (25%) over a 21-day period. However, excretion of the dose was slow and further excretion beyond 21 days would be expected, based on the long terminal elimination half-life of vandetanib (t1/2 20 days). The oral clearance rate was found to be 13.2 L/h.

7.4 Adverse effects

Rosen et al. [31] reported a significant risk of developing a rash in cancer patients receiving vandetanib. Other often reported side effects include diarrhea, tiredness, hypertension, and nausea. In one randomized phase III study of vandetanib 300 mg vs. placebo, the most prevalent side events that led to vandetanib termination were rash, pneumonia, and dyspnea. Photosensitivity responses are uncommon, but severe occurrences of vandetanib-associated photosensitivity reactions have been recorded in patients in phase I and phase II investigations. In one case, the response was severe enough that vandetanib therapy was discontinued. In addition, post-photosensitivity cutaneous hyperpigmentation was also a prevalent side effect in all cases, emphasizing the need to avoid photoexposure in vandetanib-treated patients.

References

[1] Vandetanib drugbank. Available from: https://go.drugbank.com/drugs/DB05294. Accessed Feb. 7, 2022.

[2] National Center for Biotechnology Information. PubChem Compound Summary for CID 3081361, Vandetanib.; Available from: https://pubchem.ncbi.nlm.nih.gov/compound/Vandetanib. Accessed Jan. 29, 2022.

[3] Vandetanib Theoretical Analysis. Available from: https://www.medkoo.com/products/4783. Accessed Feb. 6, 2022.

[4] Caprelsa Assessment report by EMA. Available from: https://www.ema.europa.eu/en/documents/assessment-report/caprelsa-epar-public-assessment-report_en.pdf. Accessed Feb. 11, 2022.

[5] Vandetanib chemicalbook. Available from: https://www.chemicalbook.com/ChemicalProductProperty_EN_CB01011762.htm. Accessed Feb. 22, 2022.

[6] G. Marzaro, A. Guiotto, G. Pastorini, A. Chilin, A novel approach to quinazolin-4 (3H)-one via quinazoline oxidation: an improved synthesis of 4-anilinoquinazolines, Tetrahedron 66 (4) (2010) 962–968.

[7] BL8055. *ZD6474: Determination of dissociation constants in water.*

[8] Vandetanib access data by FDA. Available from: https://www.accessdata.fda.gov/drugsatfda_docs/label/2011/022405s000lbl.pdf. Accessed Feb. 11, 2022.

[9] *ZD6474*: Determination of n-octanol/water partition coefficient. Report No. BL8076/B. Brixham Environmental Laboratory, Brixham, UK. July 2005.

[10] L.F. Hennequin, A.P. Thomas, C. Johnstone, E.S.E. Stokes, P.A. Plé, J.-J.M. Lohmann, D.J. Ogilvie, M. Dukes, S.R. Wedge, J.O. Curwen, J. Kendrew, C. Lambert-van der Brempt, Design and structure—activity relationship of a new class of potent VEGF receptor tyrosine kinase inhibitors, J. Med. Chem. 42 (26) (1999) 5369–5389.

[11] K.L. Brocklesby, J.S. Waby, C. Cawthorne, G. Smith, An alternative synthesis of Vandetanib (Caprelsa™) via a microwave accelerated Dimroth rearrangement, Tetrahedron Lett. 58 (15) (2017) 1467–1469.

[12] B. Khandare, P. Dudhe, M. Dhoke, Spectrophotometric determination of Vandetanib in bulk by area under curve and first order derivative methods, Int. J. Pharmtech Res. 12 (2019) 103–110.

[13] A.S. Abdelhameed, E.S. Hassan, M.W. Attwa, N.S. Al-Shakliah, A.M. Alanazi, H. AlRabiah, Simple and efficient spectroscopic-based univariate sequential methods for simultaneous quantitative analysis of vandetanib, dasatinib, and sorafenib in pharmaceutical preparations and biological fluids, Spectrochim. Acta A Mol. Biomol. Spectrosc. 260 (2021), 119987.

[14] A.S. Abdelhameed, M.W. Attwa, M.I. Attia, A.M. Alanazi, O.S. Alruqi, H. AlRabiah, Development of novel univariate and multivariate validated chemometric methods for the analysis of dasatinib, sorafenib, and vandetanib in pure form, dosage forms and biological fluids, Spectrochim. Acta A Mol. Biomol. Spectrosc. 264 (2022), 120336.

[15] H.W. Darwish, A.H. Bakheit, A new spectrofluorimetric assay method for vandetanib in tablets, plasma and urine, Trop. J. Pharm. Res. 15 (10) (2016) 2219–2225.

[16] H.W. Darwish, A.H. Bakheit, N.S. Al-Shakliah, I.A. Darwish, Development of novel response surface methodology-assisted micellar enhanced synchronous spectrofluorimetric method for determination of vandetanib in tablets, human plasma and urine, Spectrochim. Acta A Mol. Biomol. Spectrosc. 213 (2019) 272–280.

[17] H. Lin, D. Cui, Z. Cao, Q. Bu, Y. Xu, Y. Zhao, Validation of a high-performance liquid chromatographic ultraviolet detection method for the quantification of vandetanib in rat plasma and its application to pharmacokinetic studies, J. Cancer Res. Ther. 10 (1) (2014) 84–88.

[18] S.X. Xiang, H.-L. Wu, C. Kang, L.X. Xie, X.-L. Yin, H.-W. Gu, R.Q. Yu, Fast quantitative analysis of four tyrosine kinase inhibitors in different human plasma samples using three-way calibration-assisted liquid chromatography with diode Array detection, J. Sep. Sci. 38 (2015).

[19] B.S. Khandare, N.S. Bhujbal, S.S. Kshirsagar, Analytical method development and validation of Vandetanib by using RP-HPLC of bulk drug, Sch. Acad. J. Pharm. 8 (8) (2019) 432–435.

[20] V.L. Salode, M.D. Game, G.V. Salode, S.S. Gadge, Development of validated stability indicating method for estimation of Vandetanib and characterization of its degradants by LC-ESI-MS, Indian J. Pharm. Educ. Res. 56 (1) (2021).

[21] M. Murali, P. Venkateswara Rao, Simple, Accurate and Efficient High Performance Liquid Chromatographic Method for Determination of Vandetanib in Bulk and in Pharmaceutical Forms, 2021.

[22] J.A. Zirrolli, E.L. Bradshaw, M.E. Long, D.L. Gustafson, Rapid and sensitive LC/MS/MS analysis of the novel tyrosine kinase inhibitor ZD6474 in mouse plasma and tissues, J. Pharm. Biomed. Anal. 39 (3–4) (2005) 705–711.

[23] F. Bai, J. Johnson, F. Wang, L. Yang, A. Broniscer, C.F. Stewart, Determination of vandetanib in human plasma and cerebrospinal fluid by liquid chromatography electrospray ionization tandem mass spectrometry (LC-ESI-MS/MS), J. Chromatogr. B Analyt. Technol. Biomed. Life Sci. 879 (25) (2011) 2561–2566.

[24] I. Andriamanana, I. Gana, B. Duretz, A. Hulin, Simultaneous analysis of anticancer agents bortezomib, imatinib, nilotinib, dasatinib, erlotinib, lapatinib, sorafenib, sunitinib and vandetanib in human plasma using LC/MS/MS, J. Chromatogr. B Analyt. Technol. Biomed. Life Sci. 926 (2013) 83–91.

[25] S.M. Amer, A.A. Kadi, H.W. Darwish, M.W. Attwa, Liquid chromatography tandem mass spectrometry method for the quantification of vandetanib in human plasma and rat liver microsomes matrices: metabolic stability investigation, Chem. Cent. J. 11 (1) (2017) 017–0274.

[26] A. Denys, P. Czuczman, D. Grey, Z. Bascal, R. Whomsley, H. Kilpatrick, A.L. Lewis, Vandetanib-eluting radiopaque beads: in vivo pharmacokinetics, safety and toxicity evaluation following swine liver embolization, Theranostics 7 (8) (2017) 2164–2176.

[27] M. Attwa, A. Kadi, H. Darwish, S. Amer, N. Salem, Identification and characterization of in vivo, in vitro and reactive metabolites of vandetanib using LC–ESI–MS/MS, Chem. Cent. J. 12 (2018).

[28] S. Dall'Acqua, D. Vedaldi, A. Salvador, Isolation and structure elucidation of the main UV-A photoproducts of vandetanib, J. Pharm. Biomed. Anal. 84 (2013) 196–200.

[29] Vandetanib Monograph by BC Cancer Drug Manual. Available from: http://www.bccancer.bc.ca/drug-database-site/Drug%20Index/Vandetanib_monograph.pdf. Accessed Feb. 22, 2022.

[30] J.E. Frampton, Vandetanib, Drugs 72 (10) (2012) 1423–1436.

[31] A.C. Rosen, S. Wu, A. Damse, E. Sherman, M.E. Lacouture, Risk of rash in cancer patients treated with vandetanib: systematic review and meta-analysis, J. Clin. Endocrinol. Metabol. 97 (4) (2012) 1125–1133.

CHAPTER FIVE

Lapatinib: A comprehensive profile

Ahmed A. Abdelgalil[a] and Hamad M. Alkahtani[b]
[a]Central Laboratory, College of Pharmacy, King Saud University, Riyadh, Kingdom of Saudi Arabia
[b]Department of Pharmaceutical Chemistry, College of Pharmacy, King Saud University, Riyadh, Kingdom of Saudi Arabia

Contents

1. Physical profiles of drug substances and excipients

1.1 General information

1.1.1 Nomenclature

1.1.1.1 Systematic chemical names

N-[3-chloro-4-[(3-fluorophenyl)methoxy]phenyl]-6-[5-[(2-methylsulfonylethylamino) methyl]furan-2-yl]quinazolin-4-amine [1].

Profiles of Drug Substances, Excipients, and Related Methodology, Volume 48
ISSN 1871-5125
https://doi.org/10.1016/bs.podrm.2022.11.005
135

1.1.1.2 Nonproprietary names
Lapatinib; 231277-92-2; Lapatinib Ditosylate; GW572016 [1].

1.1.1.3 Proprietary names
Tykerb®, Tyverb®.

1.1.2 Formulae
1.1.2.1 Empirical formula, molecular weight, CAS number [1]
Empirical Formula: $C_{29}H_{26}ClFN_4O_4S$
  Molecular Weight: 581.1 g/mol
  CAS Number: 231277-92-2

1.1.2.2 Structural formula
See Fig. 1.

1.1.2.3 Simplified molecular input line entry (SMILES)
CS(=O)(=O)CCNCC1=CC=C(O1)C2=CC3=C(C=C2)
N=CN=C3NC4=CC(=C(C=C4)OCC5=CC(=CC=C5)F)Cl [1].

1.1.2.4 The IUPAC International Chemical Identifier (InChI)
InChI=1S/C29H26ClFN4O4S/c1-40(36,37)12-11-32-16-23-7-10-27
(39-23)20-5-8-26-24(14-20)29(34-18-33-26)35-22-6-9-28(25(30)15-22)
38-17-19-3-2-4-21(31)13-19/h2-10,13-15,18,32H,11-12,16-17H2,1H3,
(H,33,34,35) [1].

1.1.3 Elemental analysis
Carbon: 59.98%,
  Hydrogen: 4.14%,
  Nitrogen: 9.69%.

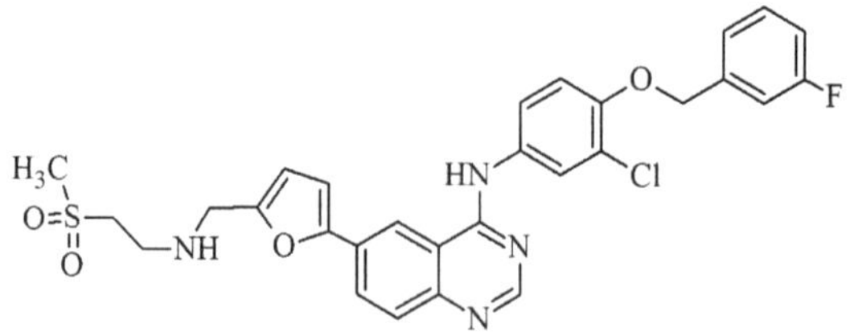

Fig. 1 Lapatinib structure.

1.1.4 Appearance
Lapatinib is pale yellow to yellow powder.

1.2 Physical characteristics
1.2.1 Ionization constants
pKa1 = 3.80; pKa2 = 7.20 [1].

1.2.2 Solubility characteristics
Very poorly soluble in water (0.007 mg/mL) and in 0.1 N HCl is 0.001 mg/mL. Soluble in DMSO at 200 mg/mL [1].

1.2.3 Partition coefficient
The octanol/water partition coefficient (LogP) value of **5.4** was reported for lapatinib [1].

1.2.4 Crystallographic properties
1.2.4.1 X-ray powder diffraction pattern
Powder X-ray diffraction of the lapatinib (free base) was evaluated by Ultima IV diffractometer (Rigaku, Japan) over the 3–140 then 3–60° 2θ range at a scan speed of 1 degree and 0.5 degree/min. The tube anode was Cu with Ka = 0.1540562 nm monochromatized with a graphite crystal. The pattern was collected at 40 kV of tube voltage and 40 mA of tube current in step scan mode (step size 0.02°, counting time 1 s per step). The X-ray powder diffraction pattern of lapatinib is given in Fig. 2 (Table 1).

1.2.5 Thermal methods of analysis
1.2.5.1 Melting behavior
The reported melting range of lapatinib is 137–139 °C [2]. Also, it has been reported that the melting point range of the pure Lapatinib base obtained is 95–98 °C (peak max. by DSC) [3].

1.2.6 Spectroscopy
1.2.6.1 UV/VIS spectroscopy
The UV absorption spectra of lapatinib (free base) in methanol solvent, is shown in Fig. 3, The spectrum was recorded using a UV-1601PC-shimadzu

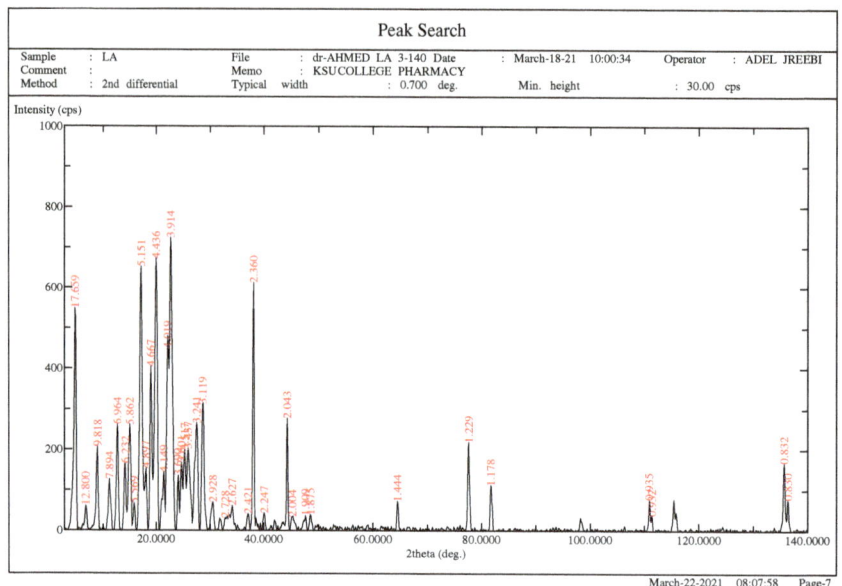

Fig. 2 X-ray powder diffraction pattern of lapatinib.

Table 1 XRD analysis of lapatinib powder (free base).

Angel (2theta)	D-value (Å)	Intensity (%)
5.000	17.6591	549
6.900	12.8002	61
9.000	9.8176	207
11.200	7.8936	127
12.700	6.9645	262
14.200	6.2320	166
15.100	5.8625	261
15.900	5.5693	67
17.200	5.1512	653
18.100	4.8970	154
19.000	4.6670	405
20.000	4.4359	673

Table 1 XRD analysis of lapatinib powder
(free base).—cont'd

Angel (2theta)	D-value (Å)	Intensity (%)
21.700	4.1487	144
22.100	4.0189	451
22.700	3.9140	724
24.100	3.6897	136
24.700	3.6014	168
25.300	3.5173	199
25.900	3.4372	202
27.500	3.2408	265
28.600	3.1186	313
30.500	2.9285	71
32.800	2.7282	32
34.100	2.6271	62
37.100	2.4213	42
38.100	2.3600	613
40.100	2.2468	43
44.300	2.0430	278
45.200	2.0044	34
47.600	1.9088	37
48.500	1.8754	40
64.500	1.4435	73
77.600	1.2293	219
81.700	1.1777	112
110.900	0.9352	78
111.400	0.9324	38
135.700	0.8317	170
136.400	0.8296	79

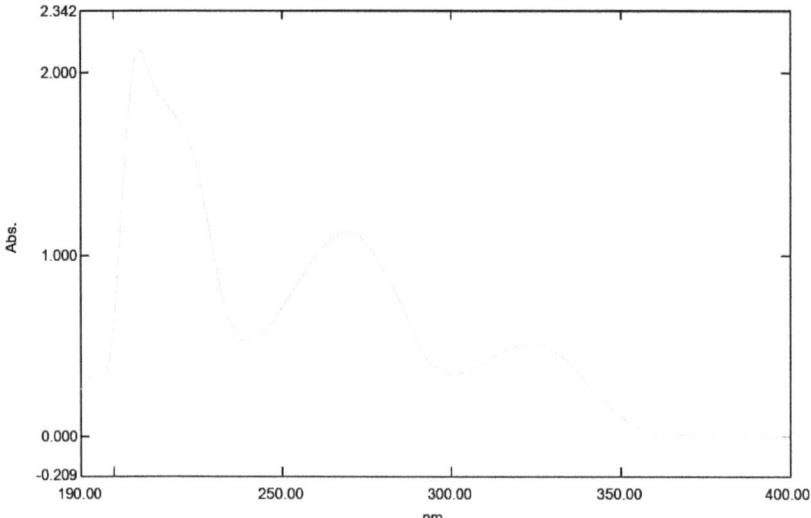

Fig. 3 UV–visible absorption spectrum of lapatinib in methanol.

double beam UV–vis spectrophotometer in 10 mm matched quartz cells. Absorption spectra were recorded from 190 to 400 nm using medium scan speed, and 2 nm Slit Width.

1.2.6.2 Vibrational spectroscopy

The infrared absorption spectrum of lapatinib was recorded as a thin film on Bruker—ALPHA II's Platinum ATR single reflection diamond ATR by applying lapatinib powder (supplied by Carbosynth Limited—UK) directly to ATR machine. The resulted IR spectrum is provided in the Fig. 4. The peaks were collected from 500 to 3500 cm^{-1}.

1.2.6.3 Nuclear magnetic resonance spectrometry

1.2.6.3.1 ^{1}H spectrum The ^{1}H spectra of Lapatinib (Shown in Fig. 5) was scanned in deuterated DMSO on Bruker 700 MHz NMR spectrometer. The parameters derived from the various NMR spectra are presented in Table 2.

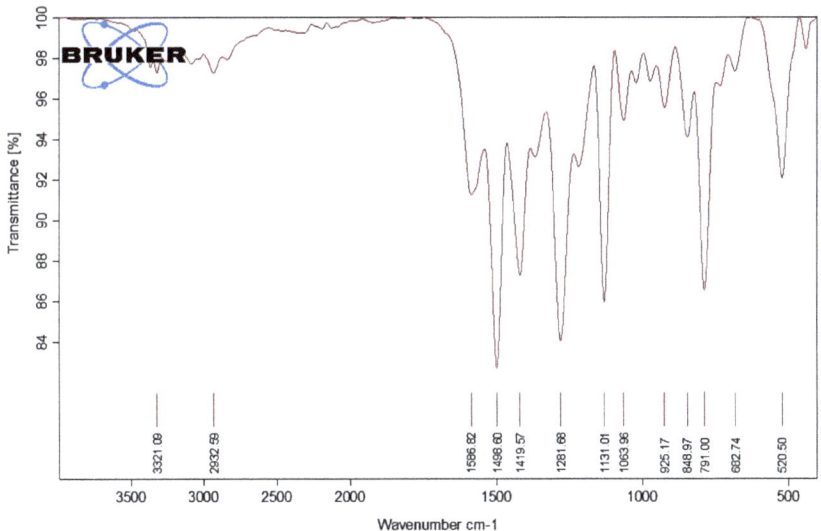

Fig. 4 Infrared absorption spectrum of lapatinib.

1.2.6.3.2 ^{13}C NMR spectrum The ^{13}C NMR spectra of crizotinib (Shown in Fig. 6) was scanned in duterated DMSO on Bruker 700 MHz NMR spectrometer. The parameters derived from the various NMR spectra are presented in Table 3.

1.2.6.4 Mass spectrometry

The mass spectrum of lapatinib was obtained using Varian CP 3800 GC combined with 320 TQMS mass spectrometer as direct probe injection using high-purity helium as the gas carrier, at a flow rate of 1 mL/min. The source temperature of MS was set at 250 °C and the Quad temperature was at 250 °C The direct probe method initially at 90 °C (held for 2 min), then was increased to 150 °C at 30 °C min^{-1} (held for 2 min), then increased further to 300 °C at 30 °C min^{-1} for 2 min. The capillary probe volume was 1 μL approx. and the scan range was set at 50–1000 mass ranges at 70 eV electron energy with no solvent delay. Fig. 7 shows the fragmentation pattern of lapatinib. The spectrum was identified with the help of library embedded in the instrument software's.

PROTON DMSO {C:\Bruker\TOPSPIN} abari 37

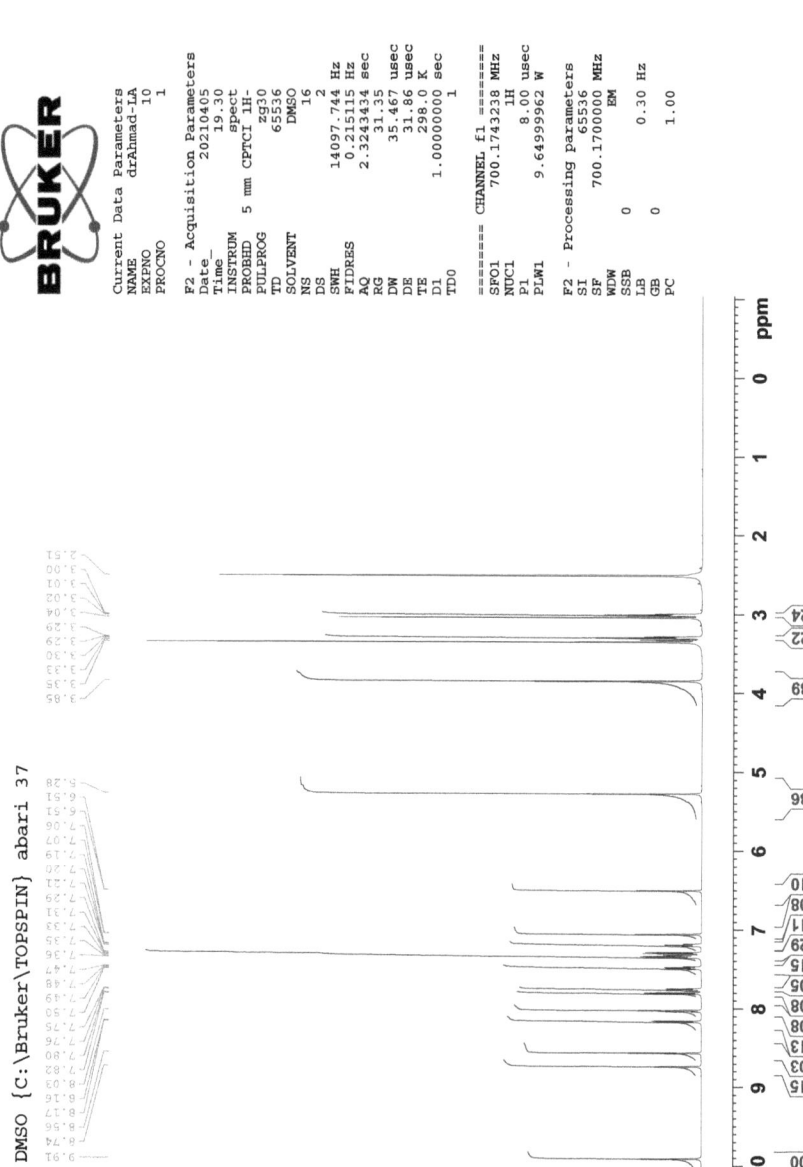

Current Data Parameters
NAME   drAhmad-LA
EXPNO   10
PROCNO   1

F2 - Acquisition Parameters
Date_   20210405
Time   19.30
INSTRUM   spect
PROBHD   5 mm CPTCI 1H-
PULPROG   zg30
TD   65536
SOLVENT   DMSO
NS   16
DS   2
SWH   14097.744 Hz
FIDRES   0.215115 Hz
AQ   2.3243434 sec
RG   31.35
DW   35.467 usec
DE   31.86 usec
TE   298.0 K
D1   1.00000000 sec
TDO   1

======== CHANNEL f1 ========
SFO1   700.1743238 MHz
NUC1   1H
P1   8.00 usec
PLW1   9.64999962 W

F2 - Processing parameters
SI   65536
SF   700.1700000 MHz
WDW   EM
SSB   0
LB   0.30 Hz
GB   0
PC   1.00

Fig. 5 ¹H NMR spectrum.

Table 2 ^{1}H NMR of lapatinib.

Signal	Chemical shift	Splitting	Integration	Assignment
3	2.99	t (J=6.58 Hz)	2	C$\underline{H}_2$NH-
1	3.03	s	3	SO$_2$C$\underline{H}_3$
2	3.28	t (J=6.58 Hz)	2	C$\underline{H}_2$SO$_2$CH$_3$
4	3.84	s	2	ArC$\underline{H}_2$NH
5	5.27	s	2	ArC$\underline{H}_2$O-
6	6.49	d (J=2.75 Hz)	1	Ar$\underline{H}$ (furan)
7	7.05	d (J=2.75 Hz)	1	Ar$\underline{H}$ (furan)
18	7.18	td (J=8.44 Hz)	1	Ar$\underline{H}$
16	7.29	d (J=8.84 Hz)	1	Ar$\underline{H}$
17	7.32	d (J=8.84 Hz)	1	Ar$\underline{H}$
19	7.34	d (J=8.44 Hz)	1	Ar$\underline{H}$
15	7.47	m	1	Ar$\underline{H}$
13	7.74	d (J=8.92 Hz)	1	Ar$\underline{H}$
8	7.80	d (J=8.68 Hz)	1	Quinazoline
12	8.01	s	1	ArN$\underline{H}$
10	8.15	d (J=8.75 Hz)	1	Quinazoline
9	8.55	s	1	Quinazoline
13	8.72	s	1	Ar$\underline{H}$
11	9.90	s	1	Quinazoline

1.3 Stability

1.3.1 Solid-state stability

Stability studies on primary and production batches of lapatinib ditosylate monohydrate [TYVERB®] revealed no significant changes could be observed and all the results remained within the specification by European Medicines Agency Evaluation of Medicines for Human Use EMEA. Further stability studies showed that lapatinib is sensitive to light, oxidation, and hydrolysis. Stability data justify the proposed re-test period when the active substance is stored at not more than 25 °C and protected from light [4].

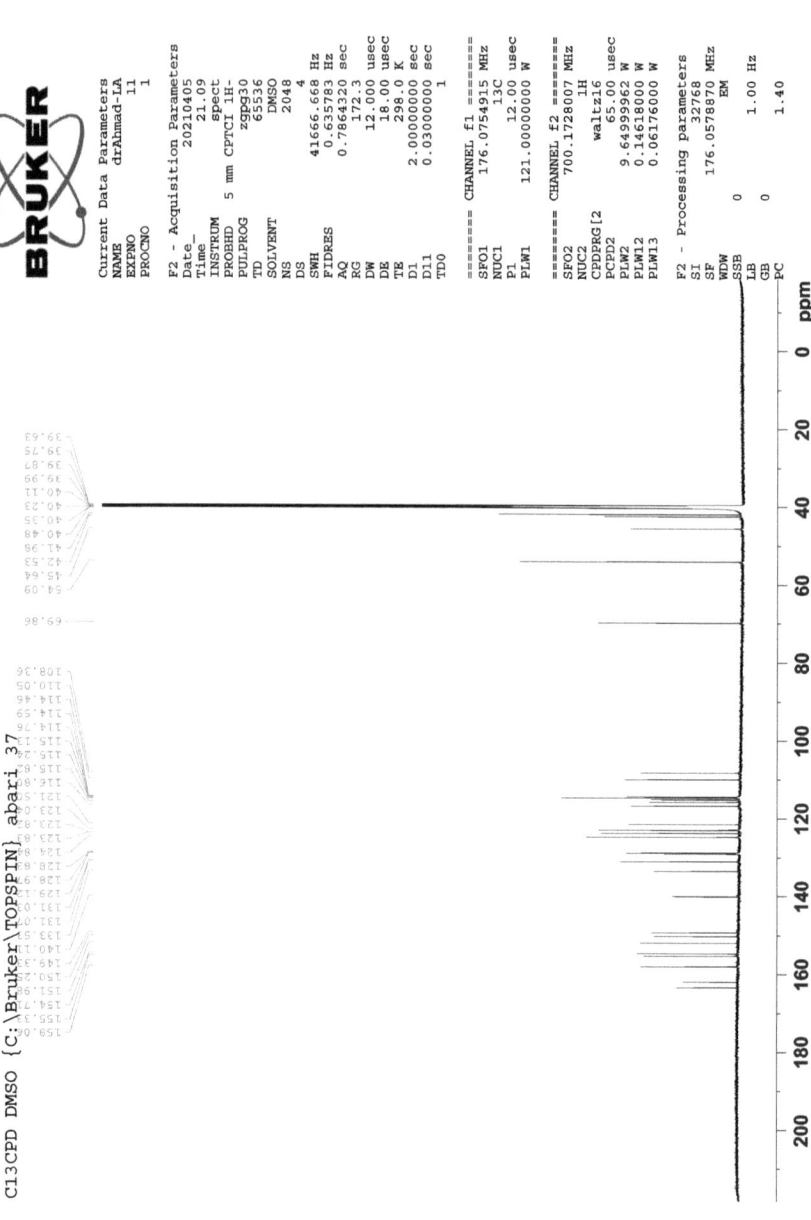

Fig. 6 ^{13}C NMR spectrum.

Table 3 ^{13}C NMR of lapatinib.

Signal	Chemical shift	Assignment
3	41.52	$\underline{C}H_2NH-$
1	42.06	$SO_2\underline{C}H_3$
4	45.17	$Ar\underline{C}H_2NH$
2	53.62	$\underline{C}H_2SO_2CH_3$
5	69.39	$Ar\underline{C}H_2O-$
6	107.89	Furan
7	109.58	Furan
25	114.06 (d, $J=21.91\,Hz$)	Ph
19	114.29	Ph
27	114.72 (d, $J=20.62\,Hz$)	Ph
23	115.35	Ph
15	116.33	Quinazoline
20	121.03	Ph
18	122.57	Ph
29	123.36 (d, $J=2.97\,Hz$)	Ph
10	124.37	Quinazoline
20	128.36	Ph
8	128.49	Quinazoline
13	128.65	Quinazoline
28	130.58 (d, $J=8.52\,Hz$)	Ph
9	133.06	Quinazoline
24	139.66 (d, $J=7.26\,Hz$)	Ph
16	148.85	Furan
21	149.78	Ph
14	151.51	Quinazoline
17	154.24	Furan
12	154.86	Quinazoline
11	157.59	Quinazoline
26	162.21 (d, $J=245.15\,Hz$)	Ph

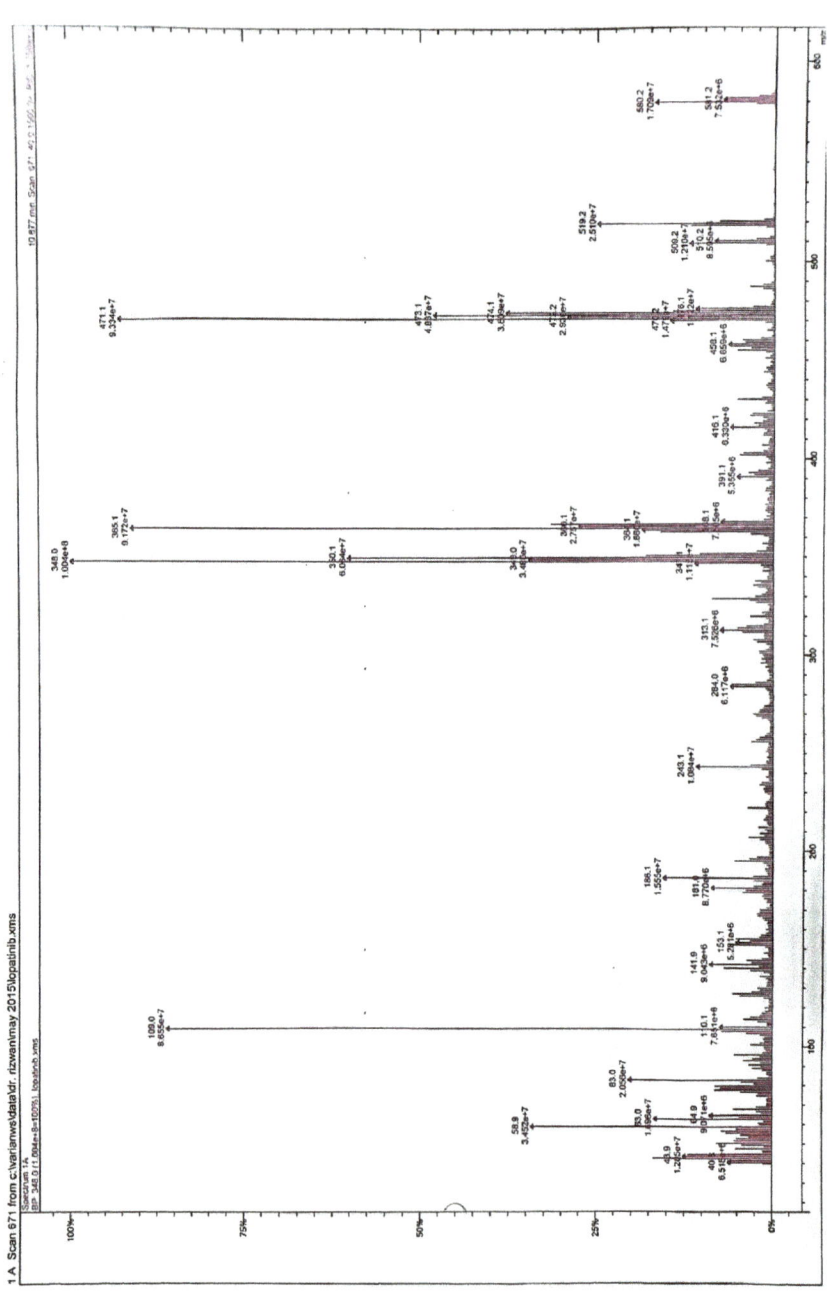

Fig. 7 Fragmentation pattern of lapatinib.

2. Analytical profiles of drug substances and excipients
2.1 Electrochemical methods of analysis
2.1.1 Voltammetry

Dogan-Topal et al. [5] reported the electro-oxidation mechanism of lapatinib (LPT) at a glassy carbon electrode (GCE) using various voltammetric techniques. The effects of pH and scan rate on LPT signal were determined in detail by cyclic voltammetry; as a result, the first and the second peaks were found diffusion and adsorption controlled, respectively. LPT exhibited three anodic peaks at 0.76 V, 0.95 V and 1.22 V by differential pulse voltammetry and two electrons were transferred for each peak. The oxidation of LPT was compared with some model compounds which contained aromatic amine structures. The possible electrooxidation pathway was also proposed. The electrochemical biosensor was designed in order to show the interaction between ct-dsDNA and LPT. For better understanding of the interaction mechanism between LPT and ct-dsDNA, spectroscopic techniques were performed. The binding constant (K) between LPT and DNA was calculated as $6.03 \times 10^5 \, M^{-1}$, $4.20 \times 10^5 \, M^{-1}$ and $3.50 \times 10^4 \, M^{-1}$ for electrochemical, UV–vis spectrophotometric and fluorescence spectroscopic techniques, respectively.

An electroanalytical method was developed for the quantification of hydrophobic lapatinib in the presence of non-ionic surfactant Triton X-100 on a glassy carbon electrode. Lapatinib presented three well-defined anodic peaks (Ep 1, Ep 2, and Ep 3) by square wave voltammetry. The oxidation behavior of Ep 1 and Ep 2 showed diffusion-adsorption mix controlled processes by cyclic voltammetry. The possible electro-oxidation mechanism is discussed. The stripping conditions and square wave voltammetry parameters were optimized. The sensitivity of the proposed method was increased in the presence of Triton X-100. The hydrophobic interaction between Triton-X-100 and lapatinib guaranteed that more drug molecules could rapidly reach the electrode surface. The adsorptive stripping square wave voltammetry exhibits a linear calibration range from 2.0×10^{-8} to $1.0 \times 10^{-6} \, mol \, L^{-1}$ for both ip 1 and ip 2 in $0.1 \, M \, H_2SO_4$ [6].

2.2 Spectroscopic methods of analysis

The fluorescence enhancing ability of kolliphor RH 40 [fluorescence enhancer surfactant] was employed in the development of a spectrofluorimetric method for the determination of lapatinib. The relative fluorescence intensity (RFI) was in linear relationship (correlation coefficient was 0.998) with the lapatinib concentrations in the range of 50–1000 ng/mL. The method limit of detection (LOD) was 27.31 ng/mL and its accuracy was $\geq$99.82% [7].

2.3 Chromatographic methods of analysis (HPLC and LCMS)

Lapatinib was quantified in various biological samples and in pharmaceutical dosage forms for different applications such therapeutic drug monitoring, stability and pharmacokinetics studies. Here is summary of lapatinib analytical methods reported in the literatures (Table 4).

2.4 Determination in body fluids and tissues

2.4.1 Enzyme-immunoassay methods

Saita et al. [37] described the production of anti-lapatinib antibody was obtained by immunizing rabbits with an antigen conjugated with bovine serum albumin using 3-chloro-4-((3-fluorobenzyl) oxy) aniline. Anti-nilotinib antibody was produced by immunizing mice with an antigen conjugated with bovine serum albumin using 2-(5-amino-2-methylanilino)-4-(3-pyridyl) pyrimidine. The generated antibodies were used to develop enzyme-linked immunosorbent assays (ELISAs) for lapatinib and nilotinib in human serum. The assays were capable of detecting lapatinib and nilotinib at serum concentrations as low as 40 and 8 ng/mL, resp.

2.4.2 Chromatographic methods

Barry et al. [38] reported method for the detection of lapatinib and its metabolites in liver tissue was determined by MSI using IR matrix-assisted laser desorption electrospray ionization (IR-MALDESI) coupled to high resolving power Fourier transform ion cyclotron resonance (FT-ICR) mass spectrometers. IR-MALDESI required minimal sample preparation while maintaining high sensitivity.

Table 4 Detailed description of reported HPLC methods for determination of lapatinib in different biological matrices and in pharmaceutical dosage forms.

No	Technique	Target	Matrix	Sample preparation	Mobile phase	Column	Detection	Linearity range	Reference
1	HPLC	Lapatinib	Bulk pharmaceuticals	Proteins precipitation with acetonitrile	Acetonitrile and water (70/30, v/v)	C18 MZ–Analytical column (5 μm, 150 × 4.6 mm, OSD-3) was used, which was protected by a (5 μm, 4.0 × 4.6 mm, OSD-3) pre-column	Detection wave length of 227 nm	5–80 μg/mL The method was validated for Accuracy and the inter- and intra-day precision were found to be 2.20%, 2.84% and 2.78%, respectively. The limits of detection and quantification were also found to be 1 and 5 μg/mL, respectively	[8]
2	HPLC	Lapatinib and erlotinib	Human plasma	Proteins precipitation with acetonitrile	Acetonitrile, methanol, water, and trifluoroacetic acid (26:26:48:0.1) at a flow rate of 1.0 mL/min	Octadecylsilyl silica gel column	UV at 316 nm	The calibration curves for lapatinib and erlotinib were in the range of 0.125–8.00 μg/mL	[9]
3	UPLC-ESI-MS/MS	Erlotinib and lapatinib	Rat plasma	Using gefitinib as the internal standard. Extracted with 4 protein precipitation (PPT) followed by solid phase extraction (SPE) using octadecyl C 18/14% cartridges	Acetonitrile (20: 80, v/v), each with 0.15% formic acid	Acquity UPLC BEH™ C18 column	Quantification was performed in the positive electrospray ionization (ESI+) mode with multiple reaction monitoring (MRM) of the transitions m/z 394.29 → 278.19 (ERL), m/z 581.07 → 365.13 (LAP), and m/z 447.08 → 128.21 gefitinib	The method was validated showing linearity over the range of 0.4–1000 (ERL) and 0.6–1000 (LAP) ng/mL with very low lower limit of quantification (LLOQ) of 0.4 and 0.6 ng/mL for erlotinib and lapatinib, respectively	[10]

Continued

Table 4 Detailed description of reported HPLC methods for determination of lapatinib in different biological matrices and in pharmaceutical dosage forms.—cont'd

No	Technique	Target	Matrix	Sample preparation	Mobile phase	Column	Detection	Linearity range	Reference
4	HPLC	Simultaneous assay of paclitaxel (PTX) and lapatinib (LPT)	Polymeric micelle formulation		Acetonitrile and water (70/30; V/V) with a flow rate of 0.5 mL/min	C18 MZ-Analytical Column (5 μm, 150 × 4.6 mm, OSD-3) which was protected with the C18 pre-column (5 μm, 4.0 × 4.6 mm, OSD-3)	227 nm	Accuracy was found to be less than 6.8%. The interday assay was evaluated to be 3.22% and 5.76% RSD for PTX and LPT, respectively. The intraday precision was found to be at its maximum value of 5.83% RSD. The limit of detection and limit of quantitation for PTX and LPT was found to be 1 μg/mL and 5 μg/mL respectively. Linearity range was 5–80 μg/mL	[11]
5	LC–MS/MS	Quantitative analysis of lapatinib	Human plasma	Exhaustive extraction with organic solvent at a flow rate of 0.2 mL/min. Running isocratically	Methanol and 0.45% formic acid in water (50:50, v/v)	XTerra C8 column (3.5 μm, 50 mm × 2.1 mm i.d.) when zileuton was used as the internal standard or on a Waters XBridge C18 column (3.5 μm, 50 mm × 2.1 mm i.d.) when lapatinib-d3 was used as the internal standard	MRM transitions for lapatinib and the internal standard (zileuton or lapatinib–d3)	The calibration curve range was 5–5000 ng/mL of lapatinib in plasma. Both the non-isotope-labeled (zileuton) and isotope-labeled (lapatinib-d3) internal standard methods showed acceptable specificity, accuracy (within 100 ± 10%), and precision (<11%)	[12]
6	Reverse-phase high-performance liquid chromatography	Lapatinib	Tablet	Gemcitabine hydrochloride as an internal standard	Mixture of acetonitrile and water (50:50v/v) as the mobile phase at a flow rate of 1.0 mL/min	ODS C-18 RP column (4.6 mm i.d. ×250 mm)	232 nm	2–60 μg/mL of lapatinib The limit of detection and limit of quantitation were 0.265 and 0.884 μg/mL, respectively	[13]

7	HPLC	Lapatinib	Human plasma	Protein precipitation with acetonitrile, IS=sorafenib	C18 Ultrabase column	Acetonitrile/20mM ammonium acetate in a proportion 53:47 (v/v) pumped at a constant flow rate of 1.2 mL/min. Isocratic elution	Ultraviolet at 260 nm	0.2–10μg/mL. Inter- and intraday coefficients of variation were less than 7%. The limit of detection and the lower limit of quantification were 0.1 and 0.2μg/mL, respectively	[14]
8	LC-MS/MS	Lapatinib	Human plasma	Liquid-liquid extraction with methyl t-butyl ether, lapatinib and isotope labeled lapatinib used as an internal standard (IS)	Zorbax SB-C18 (150×3 mm, 3.5 μm) column	An isocratic elution with the mobile phase consisting of formic buffer and the mixture of acetonitrile, methanol and formic acid was used	Mass spectrometry with positive electrospray ionization in a single ion monitoring mode was applied	The lower limit of quantification was 5 ng/mL	[15]
9	LC-MS/MS	Simultaneous determination of bortezomib (BORT), dasatinib (DASA), imatinib (IMAT), nilotinib (NILO), erlotinib (ERLO), lapatinib (LAPA), sorafenib (SORA), sunitinib (SUNI) and vandetanib (VAND)	Human plasma	Protein precipitation	Hypersil Gold® PFP column	Gradient elution of 10mM ammonium formate containing 0.1% formic acid (A) and acetonitrile containing 0.1% formic acid (B) at a flow rate of 0.3 mL/min	Electrospray ionization in positive mode using a triple quadruple mass spectrometry	2–250 ng/mL for BORT, DASA and SUNI and from 50–3500 ng/mL for the others and were fitted to a 1/x weighted linear regression model. The lowest limits of quantification were 2 ng/mL for BORT, DASA and SUNI and 50 ng/mL for the other TKIs. Intra- and inter-day RSD from 3.7% to 13.8%, accuracy (from 86.8% to 113.5%)	[16]
10	LC-MS/MS	Lapatinib	Human plasma	Solid phase extraction (SPE) columns, and 6.0 μL of the reconstituted eluate	Phenomenex CuroSil-PFP 3 mu analytical column (50×2.0 mm)	Isocratic mobile phase	Positive multiple reaction monitoring mode (m/z 581 (precursor ion) to m/z 364 (product ion) for lapatinib)	100–10,000 ng/mL, the mean recovery for lapatinib was 75% with a lower limit of quantification of 15 ng/mL	[17]

Continued

Table 4 Detailed description of reported HPLC methods for determination of lapatinib in different biological matrices and in pharmaceutical dosage forms.—cont'd

No	Technique	Target	Matrix	Sample preparation	Mobile phase	Column	Detection	Linearity range	Reference
11	HPLC-MS/MS	Simultaneous determination of dasatinib, erlotinib, gefitinib, imatinib, lapatinib, nilotinib, sorafenib and sunitinib	Human plasma	Plasma proteins were precipitated with acetonitrile using stable isotopically labeled compounds of the eight different TKIs were used as internal standards	Gradient elution	Gemini C18 column (50×2.0 mm i.d., $5.0 \mu m$ particle size)	Positive multiple reaction monitoring mode	20.0–10,000 ng/mL for erlotinib, gefitinib, imatinib, lapatinib, nilotinib and sorafenib, and from 5.00 to 2500 ng/mL for dasatinib and sunitinib. The method was validated for intra– and inter-assay accuracy ($<13.1\%$) and precision (10.0%) for all analytes	[18]
12	LC-MS/MS	Quantify cellular levels of the tyrosine kinase inhibitors dasatinib and lapatinib	Cellular samples	Extracted with a tert-butyl methyl ether: acetonitrile (3:1, v/ v):1 M ammonium formate pH 3.5 (8:1, v/v) mixture	Isocratic elution using a mobile phase of acetonitrile–10 mM ammonium formate, pH 4 (54:46, v/v), at a flow rate of 0.2 mL/min	Hyperclone BDS C18 (150 mm $\times 2.0$ mm $3 \mu m$) column	Electro-spray ionization–triple quadrupole mass spectrometry by selected reaction monitoring detection using the positive mode	The limit of detection and limit of quantification for lapatinib was determined to be 15 and 31 pg on column, respectively. The limit of detection and quantification for dasatinib was 3 and 15 pg on column, respectively	[19]
13	LC-MS/MS and Reverse–phase	Simultaneous determination of the six major TKIs	Human plasma	Protein precipitation the supernatant is diluted in ammonium formate 20 mM (pH 4.0) 1:2	Gradient elution of 20 mM ammonium formate pH 2.2 and acetonitrile containing 1% formic acid. Analyte		Electro-spray ionization–triple quadrupole mass spectrometry by selected reaction monitoring detection using the positive mode	Inter-day (CV%: 1.3–9.4%), accurate (−9.2 to +9.9%) and sensitive (lower limits of quantification comprised between 1 and 10 ng/mL)	[20]

No.	Technique	Analytes / purpose	Matrix	Sample preparation	Column	Mobile phase	Ionization mode	Performance	Ref
14	Reversed-phase chromatography coupled with tandem mass spectrometry	To facilitate therapeutic drug monitoring (TDM) for 10 anticancer compounds (dasatinib, erlotinib, gefitinib, imatinib, lapatinib, nilotinib, pazopanib, sorafenib, sunitinib, and vemurafenib) and the active metabolite, N-desethyl-sunitinib	Human plasma	Protein precipitation Stable isotopically labeled compounds were used as internal standards pretreatment with acetonitrile			The positive ion mode using multiple reaction monitoring for analyte quantification	5.00–100 ng/mL for dasatinib, sunitinib, and N-desethyl-sunitinib; 50.0–1000 ng/mL for gefitinib and lapatinib; 125–2500 ng/mL for erlotinib, imatinib, and nilotinib; and 500–10,000 ng/mL for pazopanib, sorafenib, and vemurafenib. Accuracy (bias <6.0%) and precision (12.2%) for all analytes	[21]
15	LC-MS/MS	Imatinib, dasatinib, ibrutinib, ponatinib, trametinib, sunitinib, cobimetinib, dabrafenib, erlotinib, lapatinib, nilotinib, bosutinib, sorafenib, and vemurafenib	Human plasma	Protein precipitation		Gradient elution of 10 mmol/L formate ammonium buffer containing 0.1% (vol/vol) formic acid (phase A) and acetonitrile with 0.1% (vol/vol) formic acid (phase B) at a flow rate of 300 µL/min	Positive ionization mode	The calibration curves were linear over the range from 1 to 500 ng/mL for bosutinib, cobimetinib, dasatinib, ibrutinib, and trametinib, from 5 to 500 ng/mL for ponatinib and sunitinib; from 50 to 2500 ng/mL for lapatinib; from 750 to 100,000 ng/mL for vemurafenib, and from 10 to 2500 ng/mL for dabrafenib, erlotinib, imatinib, nilotinib, and sorafenib, with coefficients of correlation above 0.99 for all analytes. The intra- and interday imprecisions were below 14.36%	[22]
16	LC-MS/MS	Imatinib, N-desmethylimatinib, dasatinib, nilotinib, erlotinib, gefitinib, lapatinib, sorafenib, sunitinib	Human plasma or serum	The internal standard solution was (150 µL imatinib-D(8), gefitinib-D(8), sunitinib-D(10), and nilotinib-(13)C (2) (15) N(2) in acetonitrile) and, after centrifugation 100 µL supernatant	50 × 0.5-mm Cyclone TurboFlow column. Analytes were focused onto a 50 × 2.1-mm (3 µm) Hypersil GOLD analytical column	Acetonitrile/water gradient	Selected reaction monitoring mode (positive APCI)	Calibration was linear (R (2) > 0.99) for all analytes. Inter- and intra-assay precision (in percent relative standard deviation, RSD) was <11% and accuracy 89–117% for all analytes	[23]

Continued

Table 4 Detailed description of reported HPLC methods for determination of lapatinib in different biological matrices and in pharmaceutical dosage forms.—cont'd

No	Technique	Target	Matrix	Sample preparation	Mobile phase	Column	Detection	Linearity range	Reference
17	UPLC/MS-MS	Imatinib, its metabolite, nilotinib, lapatinib, erlotinib and sorafenib, dasatinib, axitinib, gefitinib and sunitinib	Human plasma	Solid phase extraction (Oasis MCX µElution)	Gradient of ammonium formate-acetonitrile	BEH C18—50 × 2.1 mm column		10–5000 ng/mL for imatinib, its metabolite, nilotinib, lapatinib, erlotinib and sorafenib and from 0.1 to 200 ng/mL for dasatinib, axitinib, gefitinib and sunitinib. The mean relative extraction recovery was in the range of 90.3–106.5% thanks to the use of stable isotopes as internal standard. There was no significant ion suppression observed at the resp	[24]
18	UPLC	Lapatinib	A bulk and tablet dosage form		A mixture of 0.1% OPA buffer 300 mL (30%) and 700 mL Acetonitrile (70%) at isocratic mode	BHEL UPLC Column	309 nm using PDA detector	10–50 µg/mL for Lapatinib. Limit of detection (LOD) was 0.06 µg/mL and limit of quantification (LOQ) was 0.18 µg/mL for	[25]
19	Micellar liquid chromatography	Erlotinib, imatinib, sunitinib, sorafenib and lapatinib	Human plasma	The samples were diluted in a micellar solution and directly injected, thus clean-up steps were not required	0.13 M SDS-4% 1-butanol, buffered at pH 3.5, running under isocratic mode at 1 mL/min	C18 column	UV–visible absorbance, using a wavelength program to maximize the signal-to-noise ratio	0.05–5 µg/mol, linearity ($r^2 > 0.990$), limit of detection (15–35 ng/mL), carry-over effect, accuracy (−10.4 to +11.0%), precision (<9.2%), matrix effect, robustness (<8.4%) and stability	[26]
20	LC-MS-MS	Lapatinib	Human plasma	Protein precipitation	Isocratic elution	Zorbax SB-C18 (5 µm, 2.1 × 50 mm) column		2.50–1000 ng/mL	[27]

No.	Technique	Analyte	Matrix	Method	Mobile phase	Column	Detection	Remarks	Ref.
21	LC–MS/MS	Lapatinib	Human plasma	Liquid Liquid Extraction by 50 µL 1 mM NaOH and 2.5 mL of Et Acetate and analyzed using a reversed phase isocratic elution Pioglitazone as internal standard	Acetonitrile: 5 mM ammonium formate (80:20%volume/volume) pH adjusted to 3.80 using formic acid	Kromacil 100 C18 (4.6 × 50 mM, 5 µm) column		The lower limit of quantification is 15.004 ng/mL for lapatinib. The calibration curves are consistently accurate and precise over the concentration range of 15.004–2000.540 ng/mL in plasma for Lapatinib	[28]
22	Micellar liquid chromatography	Axitinib, lapatinib and afatinib	Human plasma		Aqueous solution of 0.07 M SDS—6.0% 1-pentanol, buffered at pH 7 with 0.01 M phosphate salt running under isocratic mode at 1 mL/min	C18 column	Absorbance at 260 nm	Linearity ($R^2 > 0.9995$), calibration range (0.5–10 mg/L), limit of detection (0.2 mg/L), carry-over effect, accuracy (−8.1 to +6.9%), precision (<13.8%)	[29]
23	RP-HPLC	Estimation of Lapatinib and its related substances	bulk and finished dosage forms		pH 4.5 ammonium formate buffer and acetonitrile as mobile phase in gradient elution mode	Zorbax Eclipse C18 (3.5 µm, 100 × 4.6 mm) column	UV detection was carried out at a wavelength of 261 nm	The LOD values were 0.009, 0.012 and 0.011 µg/mL and the LOQ values were 0.027, 0.035 and 0.0304 µg/mL resp. for impurity-1, impurity-2 and impurity-3 of Lapatinib. The average recovery values of Lapatinib related substances were found to be in the range of 97.5–101.2%. The developed method was linear over a range of 0.027–1.5 µg/mL for Lapatinib impurities	[30]

Continued

Table 4 Detailed description of reported HPLC methods for determination of lapatinib in different biological matrices and in pharmaceutical dosage forms.—cont'd

No	Technique	Target	Matrix	Sample preparation	Mobile phase	Column	Detection	Linearity range	Reference
24	UPLC-MS	Quantifiction of 18 KIs	Plasma	Solvent precipitation with acidified methanol Four deuterated internal standards (d6-erlotinib, d3-O-methyl-gefitinib, d7-lapatinib and d3-sorafenib)	Gradient elution at a flow rate of 0.25 mL/min. Initial conditions were 80% ammonium formate (10mM) with 10% acetonitrile (mobile phase A) and 20% ammonium formate (10mM) in 90% acetonitrile (mobile phase B). The proportion of mobile phase B was increased to 70% over 11 min, then reconditioned to initial conditions over 0.5 min	ACQUITY T3 HSS C18 anal. Column (150 × 2.1 mm, 1.8 μm particle size)	Time-of-flight mass spectrometry. Time-of-flight data were collected in wide pass mode, with selected ion (pseudo-MRM) spectra extracted at the precursor m/z of analytes in ESI+ mode	Coefficients of determination (r2) were invariably greater than 0.99., %RSD; for intra-day and inter-day variability was less than 10 and 17%	[31]
25	LC-MS/MS	Develop a method of high-throughput system in the analysis of imatinib (IMA), nilotinib (NIL), and lapatinib (LAP)	Human plasma	Protein precipitation with methanol containing deuterated internal standards (online solid phase extraction–MS/MS (SPE-MS/MS) system)				50–5000 ng/mL for nilotinib and imatinib, 100–10,000 ng/mL for lapatinib Intraday and interday inaccuracies within 15% and a coefficient of variation less than 15%	[32]

No.	Technique	Analytes	Matrix	Extraction	Mobile phase/Gradient	Column	Detection	Notes	Ref.
26	LC–MS/MS	Quantification of afatinib, axitinib, bosutinib, crizotinib, dabrafenib, dasatinib, erlotinib, gefitinib, imatinib, lapatinib, nilotinib, ponatinib, regorafenib, regorafenib M2, regorafenib M5, ruxolitinib, sorafenib, sunitinib, vandetanib	Human plasma	Solid phase extraction	Gradient system	C18 column (5×2.1 mm, 1.6 μm)	Ions were detected with a triple quadrupole mass spectrometry system	Drugs were arranged in four groups, according to their plasma concentration range: 0.1–200 ng/mL, 1–200 ng/mL, 4–800 ng/mL and 25–5000 ng/mL	[33]
27	LC–MS/MS	Gefitinib, erlotinib, icotinib, crizotinib, lapatinib and apatinib	Human plasma	Liquid–liquid extraction (ethyl acetate: tert-Butyl methyl ether, 1:1 v/v)	0.1% Formic acid (A) and methanol (B) with a flow rate of 0.4 mL/min. The gradient elution program was 20% B from 0 to 0.5 min, 20–95% B from 0.5 to 3.0 min, 95% B from 3.0 to 6.0 min, 95–20% B from 6.0 to 6.1 min, and 20% B from 6.1–8.0 min on gradient elution	Hypersil GOLD-C18 column (50×2.1 mm, 5 μm, Thermo Scientific)	Multiple reaction monitoring + MRM (m/z): 398.1975/212.0818 for apatinib, 450.1258/260.1506 for crizotinib, 394.1761/336.1343 for erlotinib, 447.1593/128.1070 for gefitinib, 392.1604/304.1081 for icotinib, 581.1420/458.1066 for lapatinib and 494.2662/394.1662 for imatinib	The lower limit of quantification was 0.02 ng/mL for apatinib, 0.1 ng/mL for crizotinib, 2.0 ng/mL for lapatinib and 0.05 ng/mL for erlotinib, gefitinib and icotinib	[34]

Continued

Table 4 Detailed description of reported HPLC methods for determination of lapatinib in different biological matrices and in pharmaceutical dosage forms.—cont'd

No	Technique	Target	Matrix	Sample preparation	Mobile phase	Column	Detection	Linearity range	Reference
28	LC–MS/MS	Erlotinib, sunitinib, pazopanib, axitinib, sorafenib, dasatinib, lapatinib, and nilotinib	Human plasma	Solid phase extraction	Gradient elution of acetonitrile/0.1% formic acid in water	AcQuity UHPLC BEH C18 column (50 mm × 2.1 mm ID, 1.7 μm)	An API 3200 Qtrap mass spectrometer via selective reaction monitoring operated under a pos. Scanning mode	The method was validated over a linear range of 3.13–800 nM for erlotinib; 6.25–1600 nM for sunitinib, pazopanib, and axitinib; and 12.5–3200 nM for sorafenib, dasatinib, lapatinib, and nilotinib, resp. The intra–day and inter–day precision were <16.7% for quality control samples of the analytes at the low concentration level and <13.7% for all other concentrations. The accuracy (bias) for all analytes at three different concentration levels ranged from −12.2% to 15.0%	[35]
29	Supercritical fluid chromatog. SFC–MS	Determination of 11 tyrosine kinase inhibitors (TKIs) including lapatinib	Capsules		Gradient program mobile phase A: supercritical CO_2; mobile phase B: 0.1% ammonium hydroxide in MeOH	Acquity UPC2 Torus DIOL column (3.0 mm × 100 mm, 1.7 μm)	Positive mode using Xevo G2-XS QTOF after SFC	The concentration range of 1–600 μg mL–1 for all TKIs with a coefficient of determination (R2) greater than 0.9995	[36]

3. ADME profiles of drug substances and excipients

3.1 Uses and applications

Lapatinib is a small molecule and a member of the 4-anilinoquinazoline class of kinase inhibitors, which dually inhibits the growth factor receptors ErbB1 (epidermal growth factor receptor, EGFR) and ErbB2 (HER2). It is approved by the FDA in 2007 for the use in patients with advanced metastatic breast cancer in conjunction with the chemotherapy drug capecitabine or with letrozole for the treatment of postmenopausal women with hormone receptor-positive metastatic breast cancer [39].

The recommended dosage for advanced or metastatic breast cancer is 1250 mg (5 tablets), orally once daily for 21 days in combination with capecitabine 2000 mg/m^2/day (administered orally in 2 doses approximately 12 h apart) on Days 1–14 in a repeating 21-day cycle. The recommended dose for hormone receptor-positive, HER2-positive metastatic breast cancer is 1500 mg (6 tablets) given orally once daily continuously in combination with letrozole 2.5 mg once daily. The lapatinib should be taken at least 1 h before or 1 h after a meal [39].

3.2 Absorption

Absorption following oral administration of lapatinib is incomplete and variable. Peak plasma concentrations (Cmax) of lapatinib are achieved approximately 3 h after administration. Steady state reached within 6–7 days following daily dosing of lapatinib [40]. The systemic exposure to lapatinib was increased with food, particularly with high-fat meal [41,42].

3.3 Distribution

Lapatinib is highly bound (greater than 99%) to albumin and alpha-1 acid glycoprotein. In vitro studies have indicate that lapatinib is a substrate for and inhibitor of the transporters breast cancer-resistance protein (BCRP, ABCG2) and P-glycoprotein (P-gp, ABCB1). The volume of distribution (Vd/L) of the terminal phase of lapatinib was >2200 L [42].

3.4 Metabolism

Lapatinib undergoes extensive metabolism, primarily by the cytochrome P450 (CYP) 3A4, 3A5, 2C19, and 2C8 isozymes [42].

3.5 Elimination

Lapatinib is eliminated predominantly through metabolism by CYP3A4/5 with negligible renal excretion. At clinical doses, the terminal phase half-life following a single dose was 14.2 h. Recovery of parent lapatinib in feces ranging from 3% to 67% of an oral dose [43].

3.6 Pharmacological effects

Lapatinib is an orally administered small-molecule inhibitor that targets both Epidermal Growth Factor Receptor (EGFR) and HER2. The HER, or ErbB, family of receptors contains 4 known members, HER1, HER2, HER3, and HER4. The increased activity or overexpression of HER1 (also called EGFR) and HER2 is reported to be associated with breast, lung, and colon cancers [42].

Lapatinib targets the tyrosine kinase domain of these receptors by reversibly binding to the adenosine triphosphate binding site of the kinase. This interaction prevents the phosphorylation and subsequent signal transduction of both the Ras/Raf mitogen-activated protein kinase and the phosphoinositol-3-kinase/Akt pathways, leading to an increase in apoptosis and decreased cellular proliferation. Lapatinib is able to inhibit truncated forms (lack the extracellular binding domain) of HER2 receptors which collectively known as p95HER2 receptors [42].

3.7 Use in special populations

Lapatinib administered to pregnant animals during the period of organogenesis, lapatinib caused fetal anomalies (rats) or abortions (rabbits) at maternally toxic doses, also led to death of offspring within the first 4 days after birth [43]. In patient with moderate and severe preexisting hepatic dysfunction, the systemic exposure (AUC) to lapatinib increased approximately 14% and 63%, respectively after a single oral 100-mg dose, so that lapatinib should be discontinued in such patient [43].

3.8 Drug-drug interactions

Lapatinib increased the systemic exposure (AUC) of orally or intravenously administered midazolam (CYP3A4 substrate) by 45% or 14% respectively

[44]. In cancer patients receiving lapatinib, the systemic exposure (AUC) of paclitaxel (CYP2C8 and P-gp substrate) was increased by 23% [43]. Lapatinib [1500 mg/day] increased digoxin absorption approximately 80%, implicating lapatinib inhibition of intestinal ABCB1-mediated efflux. Moreover, lapatinib was found to increase the systemic AUC of digoxin by approximately 2.8-fold [43,45].

Regarding the effect of other drugs on lapatinib, the ketoconazole was found to increase the systemic exposure (AUC) of lapatinib to approximately 3.6-fold in healthy subjects. The carbamazepine [CYP3A4 inducer] decreased the systemic exposure (AUC) to lapatinib approximately 72%, in healthy subjects receiving daily doses of carbamazepine [43]. Studies indicated that the vorinostat [suberoylanilide hydroxamic acid], which is an anticancer drug, influences the pharmacokinetic profile of lapatinib in rats, while lapatinib has no effect on vorinostat [46].

4. Methods of chemical synthesis
4.1 Preparative chemical methods

Patent WO99/35146, disclosed the process for the preparation of Lapatinib (see Scheme 1) of formula-(1), starting with 4-chloro-6-iodo-quinazoline of formula-(2), which reacted with 3-chloro-4-(3′-fluoro-benzyloxy)-aniline yielding N-[3-chloro-4-[(3′-fiuoro-benzyloxy) phenyl]]-6-iodo-quinazoline of formula-(3). Then, compound of the formula-(3) reacts with (1,3-dioxolan-2-yl)-2-(tributylstannyl)furan to get the compound of formula-(4a) which on reaction with HCl, removes the protecting group and liberates 5-(4-{3-chloro-4-(3-fluoro-benzyloxy)anilino}-6-quinazolinyl)-furan-2-carbaldehyde formula-(4). The compound of the formula-(4) on reaction with 2-methanesulfonylethylamine, followed by reduction using sodium (triacetoxy)borohydride as the reducing agent gives the required compound Lapatinib of formula-(1) as an organic residue, which is purified by column chromatography [47].

An improved process for the preparation of Lapatinib and its pharmaceutically salts also provided (see Scheme 2). This process involves, reacting 2-aminobenzonitrile of formula-(6) with iodine or iodinemonochloride to get 2-amino-5-iodobenzonitrile of the formula-(7). The compound

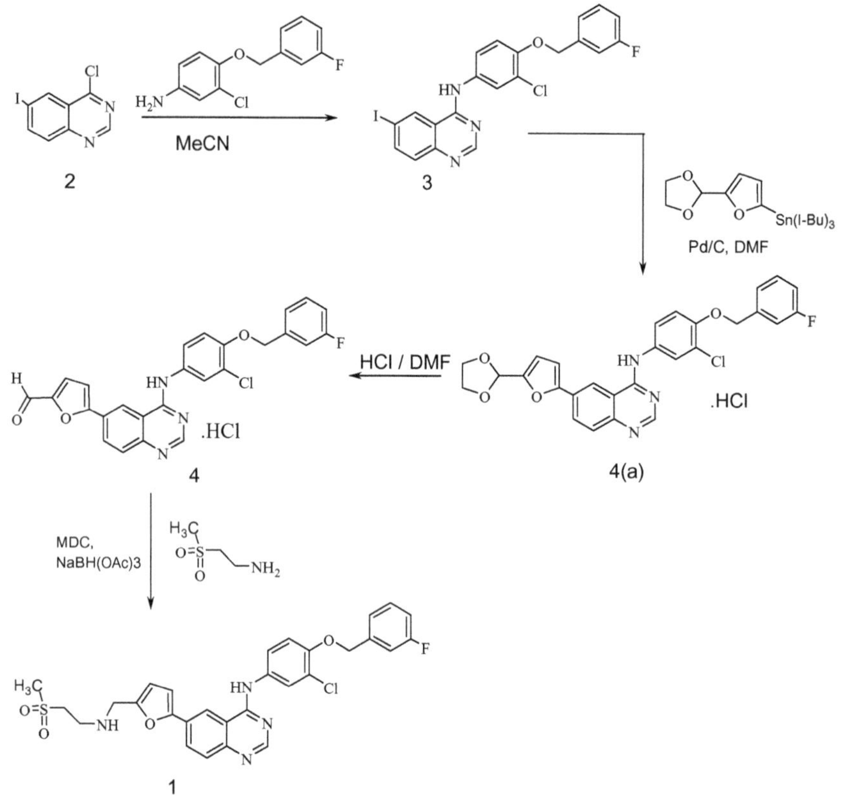

Scheme 1

of the formula-(7) on reaction with a novel compound N-(3-chloro-4-(3-fluorobenzyloxy) phenyl)-N,N-dimethylformamidine (8) at elevated temperature gives the compound N-[3-chloro-4-[(3-fluorobenzyloxy)-phenyl]-6-iodo-quinazolinamine of formula-(3), The compound of the formula-(3) on reaction with 5-formyl-2-furyl boronic acid, in presence of triethylamine and Pd/C gives the compound 5-[4-[3-chloro-4-(3-fluorobenzyloxy)anilino]-6-quinazolinyl]-furan-2-carbaldehyde of the formula-(4). The compound of the formula-(4), on reaction with 2-methanesulfonylethylamine hydrochloride gives the novel compound N [3-chloro-4[(3-fluorobenzyloxy)phenyl]-6-[5-({[2-(methanesulphonyl)-ethyl]imino}-2-furyl]-4-quinazolinamine of the formula-(9). The novel imine compound of formula-(9) on reduction, using sodium borohydride gives the compound of the formula-(1), which is Lapatinib base [47].

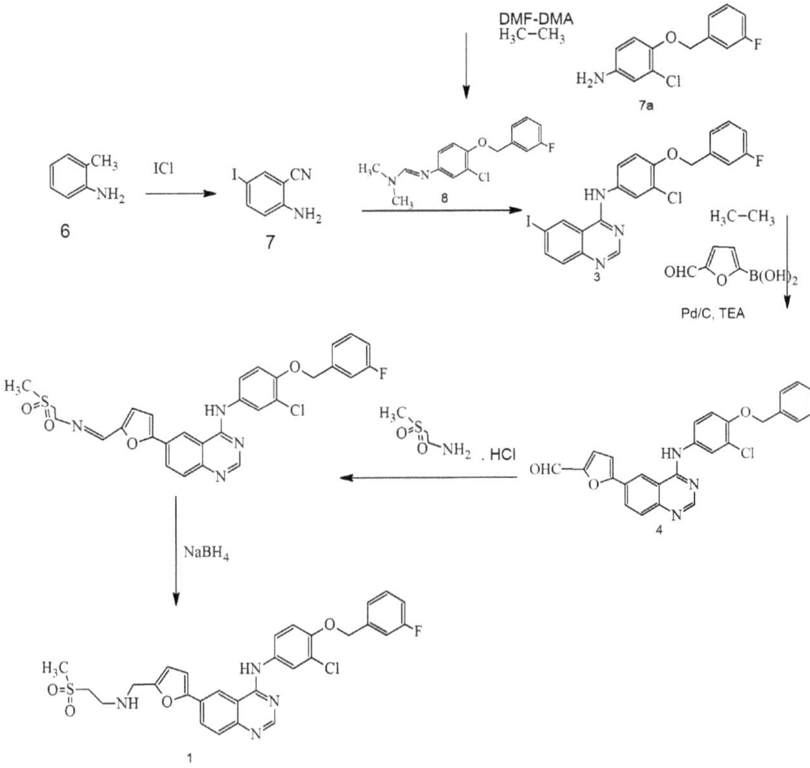

Scheme 2

References

[1] PubChem, PubChem, 2022. https://pubchem.ncbi.nlm.nih.gov/compound/208908#section=Names-and-Identifiers.

[2] https://www.scbt.com/p/lapatinib-231277-92-2.

[3] Ramanadham Jyothi Prasad APIBRAKS, Andhra Pradesh (IN); Nannapaneni Venkaiah Chowdary, Andhra Pradesh (IN) Inventor Process For The Preparation of Lapatinb and Its Pharmaceutically Acceptable Salts patent US 8,664,389 B2, 2014.

[4] G. Hamilton, B. Rath, O. Burghuber, Pharmacokinetics of crizotinib in NSCLC patients, Expert Opin. Drug Metab. Toxicol. 11 (5) (2015) 835–842.

[5] B. Dogan-Topal, B. Bozal-Palabiyik, S.A. Ozkan, B. Uslu, Investigation of anticancer drug lapatinib and its interaction with dsDNA by electrochemical and spectroscopic techniques, Sensors Actuators B Chem. 194 (2014) 185–194.

[6] Topal BEAaBD, Effect of Triton X-100 on the electrochemical behavior of hydrophobic lapatinib used in the treatment of breast cancer: A first electroanalytical study, J. Electrochem. Soc. 168 (2021) 076506.

[7] H.W. Darwish, A.H. Bakheit, N.S. Al-shakliah, A.F.M.M. Rahman, I.A. Darwish, Experimental and computational evaluation of kolliphor RH 40 as a new fluorescence enhancer in development of a micellar-based spectrofluorimetric method for determination of lapatinib in tablets and urine, PLoS One 15 (12) (2020) e0239918.

[8] E. Saadat, P. Dehghan Kelishady, F. Ravar, F. Kobarfard, F.A. Dorkoosh, Development and validation of rapid stability-indicating RP-HPLC-DAD method for the quantification of lapatinib and mass spectrometry analysis of degraded products, J. Chromatogr. Sci. 53 (6) (2015) 932–939.

[9] M. Ohgami, M. Homma, Y. Suzuki, K. Naito, M. Yamada, S. Mitsuhashi, et al., A simple high-performance liquid chromatography for determining lapatinib and erlotinib in human plasma, Ther. Drug Monit. 38 (6) (2016) 657–662.

[10] H.M. Maher, N.Z. Alzoman, S.M. Shehata, A.O. Abahussain, UPLC-ESI-MS/MS study of the effect of green tea extract on the oral bioavailability of erlotinib and lapatinib in rats: potential risk of pharmacokinetic interaction, J. Chromatogr. B Anal. Technol. Biomed. Life Sci. 1049–1050 (2017) 30–40.

[11] E. Saadat, F. Ravar, P. Dehghankelishadi, F.A. Dorkoosh, Development and validation of a rapid RP-HPLC-DAD analysis method for the simultaneous quantitation of paclitaxel and lapatinib in a polymeric micelle formulation, Sci. Pharm. 84 (2) (2016) 333–345.

[12] J. Wu, R. Wiegand, P. LoRusso, J. Li, A stable isotope-labeled internal standard is essential for correcting for the interindividual variability in the recovery of lapatinib from cancer patient plasma in quantitative LC-MS/MS analysis, J. Chromatogr. B Anal. Technol. Biomed. Life Sci. 941 (2013) 100–108.

[13] K.K. Kumar, K.E. Nagoji, R.V. Nadh, A validated RP-HPLC method for the estimation of lapatinib in tablet dosage form using gemcitabine hydrochloride as an internal standard, Indian J. Pharm. Sci. 74 (6) (2012) 580–583.

[14] V. Escudero-Ortiz, J.J. Perez-Ruixo, B. Valenzuela, Development and validation of a high-performance liquid chromatography ultraviolet method for lapatinib quantification in human plasma, Ther. Drug Monit. 35 (6) (2013) 796–802.

[15] J. Musijowski, M. Filist, P.J. Rudzki, Sensitive single quadrupole LC/MS method for determination of lapatinib in human plasma, Acta Pol. Pharm. 71 (6) (2014) 1029–1036.

[16] I. Andriamanana, I. Gana, B. Duretz, A. Hulin, Simultaneous analysis of anticancer agents bortezomib, imatinib, nilotinib, dasatinib, erlotinib, lapatinib, sorafenib, sunitinib and vandetanib in human plasma using LC/MS/MS, J. Chromatogr. B Anal. Technol. Biomed. Life Sci. 926 (2013) 83–91.

[17] F. Bai, B.B. Freeman 3rd, C.H. Fraga, M. Fouladi, C.F. Stewart, Determination of lapatinib (GW572016) in human plasma by liquid chromatography electrospray tandem mass spectrometry (LC-ESI-MS/MS), J. Chromatogr. B Anal. Technol. Biomed. Life Sci. 831 (1-2) (2006) 169–175.

[18] N.A. Lankheet, M.J. Hillebrand, H. Rosing, J.H. Schellens, J.H. Beijnen, A.D. Huitema, Method development and validation for the quantification of dasatinib, erlotinib, gefitinib, imatinib, lapatinib, nilotinib, sorafenib and sunitinib in human plasma by liquid chromatography coupled with tandem mass spectrometry, Biomed. Chromatogr. 27 (4) (2013) 466–476.

[19] S. Roche, G. McMahon, M. Clynes, R. O'Connor, Development of a high-performance liquid chromatographic-mass spectrometric method for the determination of cellular levels of the tyrosine kinase inhibitors lapatinib and dasatinib, J. Chromatogr. B Anal. Technol. Biomed. Life Sci. 877 (31) (2009) 3982–3990.

[20] A. Haouala, B. Zanolari, B. Rochat, M. Montemurro, K. Zaman, M.A. Duchosal, et al., Therapeutic drug monitoring of the new targeted anticancer agents imatinib, nilotinib, dasatinib, sunitinib, sorafenib and lapatinib by LC tandem mass spectrometry, J. Chromatogr. B Anal. Technol. Biomed. Life Sci. 877 (22) (2009) 1982–1996.

[21] M. Herbrink, N. de Vries, H. Rosing, A.D. Huitema, B. Nuijen, J.H. Schellens, et al., Quantification of 11 therapeutic kinase inhibitors in human plasma for therapeutic drug monitoring using liquid chromatography coupled with tandem mass spectrometry, Ther. Drug Monit. 38 (6) (2016) 649–656.

[22] H.H. Huynh, C. Pressiat, H. Sauvageon, I. Madelaine, P. Maslanka, C. Lebbe, et al., Development and validation of a simultaneous quantification method of 14 tyrosine kinase inhibitors in human plasma using LC-MS/MS, Ther. Drug Monit. 39 (1) (2017) 43–54.

[23] L. Couchman, M. Birch, R. Ireland, A. Corrigan, S. Wickramasinghe, D. Josephs, et al., An automated method for the measurement of a range of tyrosine kinase inhibitors in human plasma or serum using turbulent flow liquid chromatography-tandem mass spectrometry, Anal. Bioanal. Chem. 403 (6) (2012) 1685–1695.

[24] S. Bouchet, E. Chauzit, D. Ducint, N. Castaing, M. Canal-Raffin, N. Moore, et al., Simultaneous determination of nine tyrosine kinase inhibitors by 96-well solid-phase extraction and ultra performance LC/MS-MS, Clin. Chim. Acta 412 (11-12) (2011) 1060–1067.

[25] S. Biswal, S. Mondal, Analytical method validation report for assay of Lapatinib by UPLC, Pharm. Methods 10 (1) (2019) 9–14.

[26] I. Garrido-Cano, A. Garcia-Garcia, J. Peris-Vicente, E. Ochoa-Aranda, J. Esteve-Romero, A method to quantify several tyrosine kinase inhibitors in plasma by micellar liquid chromatography and validation according to the European Medicines Agency guidelines, Talanta 144 (2015) 1287–1295.

[27] G.P. Kocan, M. Huang, F. Li, S. Pai, A sensitive LC-MS-MS assay for the determination of lapatinib in human plasma in subjects with end-stage renal disease receiving hemodialysis, J. Chromatogr. B Anal. Technol. Biomed. Life Sci. 1097–1098 (2018) 74–82.

[28] P. Ranganathan, V. Gunasekaran, I. Singhvi, Bio analytical method development and validation of lapatinib in human plasma by LC—MS/MS, Eur. J. Biomed. Pharm. Sci. 4 (8) (2017) 781–789.

[29] J. Albiol-Chiva, J. Esteve-Romero, J. Peris-Vicente, Development of a method to determine axitinib, lapatinib and afatinib in plasma by micellar liquid chromatography and validation by the European Medicines Agency guidelines, J. Chromatogr. B Anal. Technol. Biomed. Life Sci. 1074–1075 (2018) 61–69.

[30] R. Ivaturi, M.T. Sastry, S. Satyaveni, Development and validation of stability indicating HPLC method for the determination of Lapatinib impurities in bulk and finished formulations, Int. J. Pharm. Sci. Res. 8 (7) (2017) 3081–3091.

[31] M. van Dyk, J.O. Miners, G. Kichenadasse, R.A. McKinnon, A. Rowland, A novel approach for the simultaneous quantification of 18 small molecule kinase inhibitors in human plasma: a platform for optimized KI dosing, J. Chromatogr. B Anal. Technol. Biomed. Life Sci. 1033–1034 (2016) 17–26.

[32] I. Vrobel, H. Janeckova, E. Faber, K. Bouchalova, K. Micova, D. Friedecky, et al., Ultrafast online SPE-MS/MS method for quantification of 3 tyrosine kinase inhibitors in human plasma, Ther. Drug Monit. 38 (4) (2016) 516–524.

[33] C. Merienne, M. Rousset, D. Ducint, N. Castaing, K. Titier, M. Molimard, et al., High throughput routine determination of 17 tyrosine kinase inhibitors by LC-MS/MS, J. Pharm. Biomed. Anal. 150 (2018) 112–120.

[34] M.W. Ni, J. Zhou, H. Li, W. Chen, H.Z. Mou, Z.G. Zheng, Simultaneous determination of six tyrosine kinase inhibitors in human plasma using HPLC-Q-Orbitrap mass spectrometry, Bioanalysis 9 (12) (2017) 925–935.

[35] Y. He, L. Zhou, S. Gao, T. Yin, Y. Tu, R. Rayford, et al., Development and validation of a sensitive LC-MS/MS method for simultaneous determination of eight tyrosine kinase inhibitors and its application in mice pharmacokinetic studies, J. Pharm. Biomed. Anal. 148 (2018) 65–72.

[36] S. Zhang, W. Jin, Y. Yang, Simultaneous identification and determination of eleven tyrosine kinase inhibitors by supercritical fluid chromatography-mass spectrometry, Anal. Methods 11 (16) (2019) 2211–2222.

[37] T. Saita, Y. Yamamoto, M. Shin, Y. Nakano, Preparation of antibodies and development of an enzyme-linked immunosorbent assay for the tyrosine kinase inhibitors
lapatinib and nilotinib, Biol. Pharm. Bull. 38 (10) (2015) 1652–1657.

[38] J.A. Barry, M.R. Groseclose, G. Robichaud, S. Castellino, D.C. Muddiman, Assessing
drug and metabolite detection in liver tissue by UV-MALDI and IR-MALDESI
mass spectrometry imaging coupled to FT-ICR MS, Int. J. Mass Spectrom. 377
(2015) 448–455.

[39] W. Shu, L. Ma, X. Hu, M. Zhang, W. Chen, W. Ma, et al., Drug-drug interaction
between crizotinib and entecavir via renal secretory transporter OCT2, Eur.
J. Pharm. Sci. 142 (2020) 105153.

[40] A.K. Bence, E.B. Anderson, M.A. Halepota, M.A. Doukas, P.A. DeSimone, G.A.
Davis, et al., Phase I pharmacokinetic studies evaluating single and multiple doses of
oral GW572016, a dual EGFR-ErbB2 inhibitor, in healthy subjects, Investig. New
Drugs 23 (1) (2005) 39–49.

[41] F. Xu, K. Lee, W. Xia, H. Liao, Q. Lu, J. Zhang, et al., Administration of lapatinib
with food increases its plasma concentration in chinese patients with metastatic breast
cancer: a prospective phase II study, Oncologist 25 (9) (2020) e1286-e91.

[42] P.J. Medina, S. Goodin, Lapatinib: a dual inhibitor of human epidermal growth factor
receptor tyrosine kinases, Clin. Ther. 30 (8) (2008) 1426–1447.

[43] FDA lapatinib package insert. https://www.novartis.us/sites/www.novartis.us/files/
tykerb.pdf.

[44] K.M. Koch, E.C. Dees, S.A. Coker, N.J. Reddy, S.D. Gainer, N. Arya, et al., The
effects of lapatinib on CYP3A metabolism of midazolam in patients with advanced
cancer, Cancer Chemother. Pharmacol. 80 (6) (2017) 1141–1146.

[45] K.M. Koch, D.A. Smith, J. Botbyl, N. Arya, L.P. Briley, L. Cartee, et al., Effect
of lapatinib on oral digoxin absorption in patients, Clin. Pharmacol. Drug Dev. 4
(6) (2015) 449–453.

[46] F. Lin, S. Wang, Y. Zhou, C. Wu, H. Zou, P. Geng, et al., Pharmacokinetic interaction
study combining lapatinib with vorinostat in rats, Pharmacology 95 (3-4) (2015)
160–165.

[47] R.J. Jyothi Prasad, A novel process for the preparation of lapatinib and its pharmaceutically acceptable salts (WIPO WO2010061400A1), 2009. https://patentimages.storage.
googleapis.com/af/b0/5c/785b69a6a6b236/WO2010061400A1.pdf.

CHAPTER SIX

Pharmaceutical based cosmetic serums

Nimra Khan[a], Sofia Ahmed[b], Muhammad Ali Sheraz[a,b], Zubair Anwar[c], and Iqbal Ahmad[c]

[a]Department of Pharmacy Practice, Baqai Institute of Pharmaceutical Sciences, Baqai Medical University, Karachi, Pakistan
[b]Department of Pharmaceutics, Baqai Institute of Pharmaceutical Sciences, Baqai Medical University, Karachi, Pakistan
[c]Department of Pharmaceutical Chemistry, Baqai Institute of Pharmaceutical Sciences, Baqai Medical University, Karachi, Pakistan

Contents

1. Introduction

Different pharmaceutical dosage forms are used for the treatment of several diseases. These may include solid dosage forms (tablets, capsules, lozenges, granules, powders, dry powder inhalers, chewable tablet, suppositories, etc.), semisolid dosage forms (ointments, creams, gels, pastes, foams, etc.), liquid dosage forms (syrups, elixirs, linctuses, drops, liniments, lotions, gargles, mouthwashes, throat paints, etc.), biphasic liquid dosage forms (suspensions, colloids, emulsions, etc.) and gaseous dosage forms (aerosols). Each dosage form possesses multiple benefits, but the selection of a particular dosage form should be made according to the patient's condition and ease of administration [1–4].

Before developing a pharmaceutical dosage form or any other formulation, the pilot study and pre-formulation testing should be carried out to collect the necessary information to establish a stable and good bioavailable dosage form [5,6]. However, two steps generally involved in the pre-formulation studies are:

Step I: Preliminary investigation or molecular optimization.

Step II: Detection and molecular modification.

Some of the factors generally considered, and parameters evaluated for pre-formulation studies are presented in Table 1.

Table 1 New drug development evaluation parameters.

Factors	Evaluation parameters
Stability	Light (UV and visible), temperature, pH, and humidity
Compatibility of the solid material	Thin-layer chromatography (TLC), differential reflectance spectroscopy (DRS), and electron spectroscopy for analysis of the chemicals
Physicochemical properties	Electrical conductivity, total dissolved and suspended solids, melting point, turbidity testing, molecular weight, and structure
Thermal stability analysis	Thermogravimetric analysis (TGA), differential thermal analysis (DTA), and differential scanning calorimetry (DSC)
Physicomechanical properties	Relative density, yield strength, hardness, toughness, fracture, flexural strength, thermal, and electrical conductivities
Absorbance spectra	Ultraviolet and visible (UV–vis) spectroscopy, infrared spectroscopy
In vitro studies	Dissolution studies and testing of solid-state drugs
Other properties	Bulk characteristics variability, rotary polarization, and polymorphism

2. Structure and function of the human skin

The skin is the largest, waterproof, and most protective external barrier between the internal structured organs. The total weight of human adult skin is approximately 3.6 kg. The structure of the skin is shown in Fig. 1. It is composed of generally three types of layers: epidermis, dermis, and hypodermis. The epidermis layer consists of a total of five portions, i.e., stratum corneum (outermost epidermis portion), stratum lucidum (present only hairless part of the body), stratum granulosum, stratum spinosum, and stratum basale (deepest epidermis portion). However, this layer contains many cells such as melanocytes, Langerhans's cells, and Merkel's cells, but keratinocytes are the most dominant epidermis cell type. Beneath the epidermis layer, a thick, fibrous, and elastic layer is present, known as the dermis which contains stromal cells. The papillary dermis (upper and thin portion) and reticular dermis (lower and thick portion) are the two portions of the dermis layer. The third deepest layer is a hypodermis layer, also known as subcutaneous tissue or panniculus, formed by fats and connective tissues [7–11].

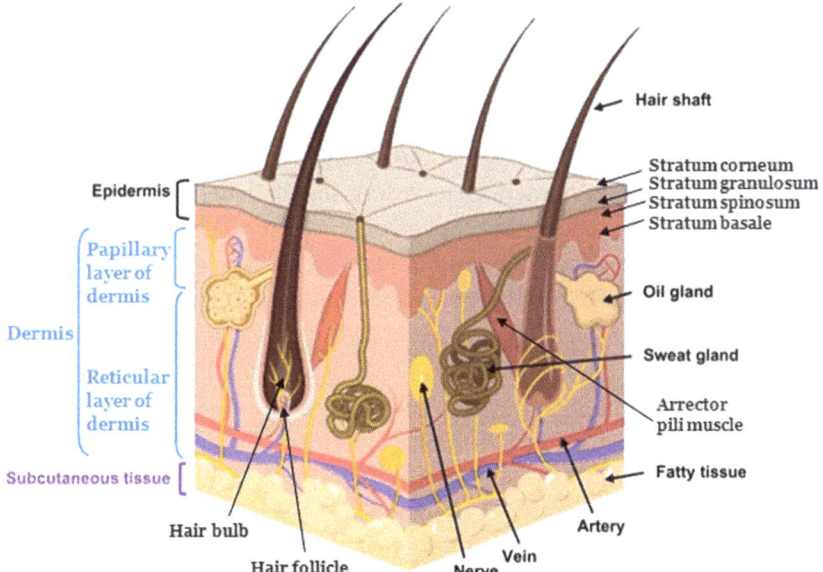

Fig. 1 Anatomy of the human skin. *Reproduced with permission from P.D. Amin, B.T. Raimi-Abraham, D.S. Shah, S. Gurram. Medicated topicals, In: Adeboye Adejare (Ed.), Remington the Science and Practice of Pharmacy. 23rd edition, Elsevier Inc., London, UK, 2021, 381–393.*

The intricate network of this largest organ provides a shielding barrier effect against the different environmental stresses. It acts as a sensory organ and plays an important role in the homeostasis process as well as in endocrine and endocrine functions. The skin generates some antimicrobial peptides (cathelicidins and β-defensins, etc.) which play a preventative role in the case of infections [10,12].

3. Serums and its types

Cosmetic dermatology also belongs to the healthcare system. The popularity of this field is gradually increasing due to several benefits such as non-invasive procedures and the development of some highly effective techniques [13–15]. According to the American Society for Dermatological Surgery (ASDS), in 2018, a lot of money (>12.5 million USD) has been spent on invasive and non-invasive procedures [16]. Topical cosmeceuticals are widely used in different skin related problems such as psoriasis, mycosis, acne, hives, erythema, dermatitis, etc., and therefore various pharmaceutical dosage forms have been designed and used for the treatment and improvement of worst skin conditions [15,17].

The term "cosmeceutical" is a combination of two industries, i.e., cosmetic and pharmaceutical [18,19]. The Food, Drug and Cosmetic (FD&C) Act and the European Union do not recognize the term "cosmeceutical" but the cosmetic industry does use this word to refer to cosmetic products that have medicinal or drug-like benefits [19,20]. The word "cosmetic" means a preparation applied on the external surface of the human for cleansing, protecting, boost up attractiveness, beautifying, and restore the physical appearance without damaging the internal body's structure and functions. However, "pharmaceutical" means a bioactive compound, manufactured for use as a medicinal drug [18,19]. Cosmeceuticals are products that combine cosmetics and bioactive compounds. Currently, cosmeceuticals are a segregated subclass within the domain of a cosmetic or drug. In Europe and Japan, cosmeceuticals are a subclass of cosmetics while in the US they are considered as a subclass of drugs [19].

Due to this confusion in nomenclature around the globe, certain things are unclear or varied from region to region as the parameters for drugs and cosmetics are different according to the local legislation. Some of those factors may include labeling rules whether the cosmeceutical is given on prescription or as OTC, testing protocols, safety evaluation, etc. [19]. Although there is no definite/official classification of cosmeceuticals, they are generally classified based on their indication and are reviewed by Pandey et al. [19]. Similarly, a topical drug classification system (TCS) is also been defined that is based on the qualitative and quantitative composition of the dosage form and the in vitro release rate of the active ingredient [21].

Pharmaceutical-based cosmetic serums have recently become one of the most popular skincare products with a very high concentration of active ingredient or bioactive compound in their formula with minimum viscosity and good spreadability [18]. The qualitative and quantitative attributes can be considered as those reported for topical cosmetic preparations or solutions for topical use in the official compendia (e.g., United States Pharmacopeia) and various guidelines issued by FDA for drugs and cosmetics (e.g., Guide to Inspection of Topical Drug Products, Cosmetics GMP, etc.). The pharmaceutical serum or pharmaceutical-based cosmetic serum is relatively a new dosage form that has been developed and widely used in many cosmetic non–invasive procedures. The high skin penetration (epidermis to the dermis) of serum has been found to produce quick and tremendous effects on the skin [17,18,22–25]. There are mainly seven types of topical serums that are used for different skin diseases and cosmetic purposes (Fig. 2). They are discussed as follows.

3.1 Facial serums

Many dermatologists and cosmetologists are prescribing serums to reverse several skin issues with a minimum duration of time. Different environmental conditions may bring about changes in the appearance of facial skin leading to poor texture, appearance of wrinkles, acne, rosacea, UV-induced dark spots, dryness, pigmentation issues (hypopigmentation and hyperpigmentation), and dullness of skin [23,26–33]. Since facial serums are highly concentrated and are lighter in weight, they are formulated to deliver a higher amount of active ingredient(s) more quickly and efficiently [17,22,25,34]. They are available in small bottles (usually 15–30 mL) with a dropper attached to the cap as only a few drops are enough to treat different facial skin problems [35,36]. There are two main classifications of facial serums depending on the type of base used (water or oil) and the type of skin to which they are to be applied. Based on the type of formulation, the facial serums can further be classified into:
- Water-based serums
- Oil-based serums

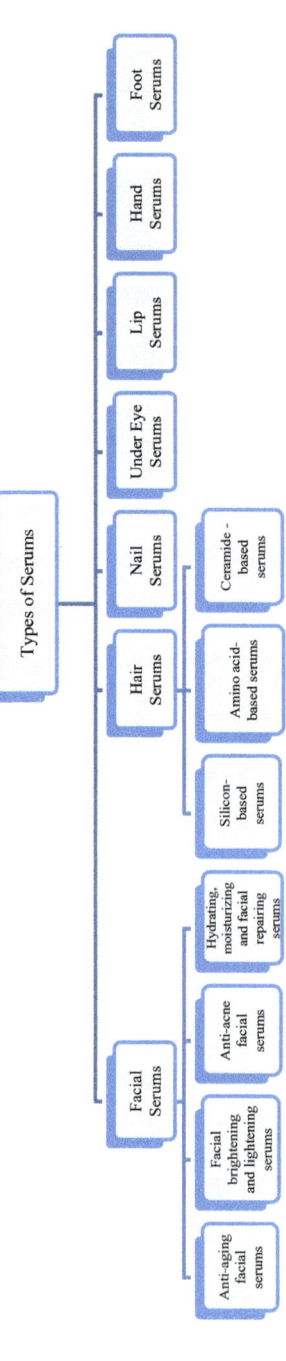

Fig. 2 Types of serums.

3.1.1 Water-based serums

This type of facial serums usually contains water or *Aloe vera* as a base and is widely used for the treatment of UV-damaged, unhealthy, dehydrated, freckle, and acne-prone skin [37–39]. In comparison with oil-based serums, water-based serums are used mainly in a humid and warm climate for controlling excess facial oil. However, they are also used in cold weather for the hydration of dry, scaly skin [22,40]. Various hyaluronic acid and vitamin C cosmetic serums are water-based and have been used to correct skin hydration as well as recover the UV-induced different skin problems such as aging and hyperpigmentation [34,41,42]. The rutin-based facial serum is another example of an aqueous phase containing cosmetic serum. Interestingly, this antioxidant-based serum gives an anti-aging effect very quickly and provides youthful and brightening skin [43]. Many marketed aqueous-based cosmetic serums such as L'Unique Miracular Facial Serum® and kinase C8 Peptide Intensive Treatment® are predominantly used for anti-aging and skin brightening purposes [29,44]. Sometimes, chemical face peelers are used to remove the outer skin (stratum corneum), and then these facial serums are applied for the rapid and deeper penetration of the active ingredient(s).

3.1.2 Oil-based serums

Oil-based serums are suitable for dry scabrous skin [45–48]. These serums are applied both in the day (before the sunscreen) and at night time according to the demand of the skin to be moisturized. They protect from UV radiation, pollution, and dry climate damages. These serums may also be used in all skin types and are also used in exceptional cases such as dermatitis, eczema, and acne blemish-prone skin [48,49]. Argan oil, jojoba oil, or avocado oil is generally used as a vehicle in these serums [18,46,50,51]. A variety of essential oils are used in the preparation of oil-based serums. These essential oils may contain one or more active constituents that produce the therapeutic effect [46]. The commonly used essential oils and their active constituents may include anise (anethole, anisaldehyde, linalool) [52], bergamot (linalyl acetate, limonene, linalool, and γ-terpinene) [53], black pepper (trans-caryophyllene, limonene, and β-pinene) [54], balsam (β-pinene, borynyl acetate, δ-3-carene) [52], clary sage (linalyl acetate and linalool) [52], clove (eugenol, β-caryophyllene and eugenyl acetate) [52], chamomile (α-pinene, terpinene-4-ol, and chamazulene) [52,55], coriander (linalool, geranyl acetate) [52], cypress (α-pinene, δ-3-carene, and limonene) [56], eucalyptus (1–8-cineol and α-pinene) [52], fennel (limonene and (*E*)-anethole) [52], grapefruit (limonene and myrcene)

[57], ginger (α-zingiberene, β-bisabolene, β-sesquiphellandrene, limonene, citral, geraniol) [58], juniper (α-pinene and myrcene) [52], jasmine (linalool and benzyl acetate) [52], lavender (linalyl acetate, linalool and terpinen-4-ol) [52], lemongrass (geranial, neral, myrcene, limonene and citral) [52], lemon myrtle (geranial and neral) [52], lime (limonene, α-terpineol, terpinen-4-ol, β-pinene, citral) [52], marjoram (terpinen-4-ol, trans-sabinene hydrate, γ-terpinene) [59], macadamia (oleic acid, palmitoleic acid, linoleic acid and α-linolenic acid) [60], mandarin (limonene and γ-terpinene) [61], pine needle (α-terpineol, linalool and, limonene) [62], pimento berries (eugenol, methyl eugenol and caryophyllene) [63], petitgrain (linalool, linalyl acetate and α-terpineol) [64], rosemary (p-cymene, linalool, γ-terpinene and thymol) [65], rosewood (1,8-cineole, α-terpinyl acetate, sabinene, 4-terpinen-4-ol, and myrcene) [66], sandalwood (α-santalol and β-santalol) [52], spearmint (carvone, cis-carveol and limonene) [67], sea fennel (γ-terpinene, β-phellandrene and sabinene) [68], spikenard (sesquiterpene and sesquiterpenols) [69], sage (α-thujone, camphor, and α-pinene) [70], thyme (carvacrol and γ-terpinene) [52], vetiver (zizanol and β-vetirenene) [52], and wintergreen (methyl salicylate) [71], respectively. However, some other popular essential oils are discussed below.

3.1.2.1 Carrot seed essential oil
The major constituents of this essential oil are carotol (66.78%), daucene (8.74%), (Z,Z)-α-farnesene (5.86%), germacrene-D (2.34%), trans-α-bergamotene (2.41%) and β-selinene (2.20%), respectively. Carotol possesses suitable anti-aging properties and decreases the size of dark spots and scars on the skin [72]. It possesses antioxidant activity, which minimizes facial skin creases, inflammation, and UV radiation-induced pigmentation problems. It also builds up the new growth of the facial cells [47,73,74].

3.1.2.2 Frankincense essential oil
The α-pinene (64.7%), α-thujene (52.4%), β-pinene (13.1%), myrcene (22.4%), sabinene (7.0%), limonene (20.4%), p-cymene (16.9%), and β-caryophyllene (10.5%) are the main components of this essential oil and gives excellent effect on all types of skin [75,76]. The α-pinene and myrcene (monoterpenes) are responsible for the recovery of the UVB-induced photoaging skin problems [77]. This oil also improves facial tone by diminishing the pore size [78,79]. It is considered best for acne-prone skin because it

contains anti-inflammatory as well as anti-bacterial properties. It also helps to stimulate the regeneration of cells and aid in the nourishment of dry and flaky skin [75].

3.1.2.3 Geranium essential oil

The citronellol (37.5%), geraniol (6.0%), caryophyllene oxide (3.7%), menthone (3.1%), linalool (3.0%), β-bourbonene (2.7%), isomenthone (2.1%), geranyl formate (2.0%), cis-rose oxide (1.9%), geranyl tiglate (1.8%), and 2-phenylethyl tiglate (1.5%) are found in geranium essential oil [80]. According to some studies, two monoterpenes, i.e., citronellol and geraniol possess anti-fungal activity against the *Trichophyton rubrum* known as ringworm and *Candida albicans* strains [52,81,82]. Due to these constituents, this essential oil has been added in some dermatological preparations to recover the issues of skin, especially facial bruises, eczema, and dermatitis [52,83,84]. It also repairs the broken capillaries and enhances the circulation of blood in the applied area. Additionally, it controls the oil imbalance with the correction of face oil production [52,84–87].

3.1.2.4 Lavender essential oil

Out of the 78 components, the main ingredients of this oil are linalool (30.6%), linalyl acetate (14.2%), geraniol (5.3%), β-caryophyllene (4.7%), and lavandulyl acetate (4.4%), respectively [88]. It works amazingly on the facial skin and reduces the size of fine lines, scars, and dark spots. Moreover, it provides a soothing effect on the dry patchy facial skin, generates new facial cells, and gives fresh and flawless skin [89–91].

3.1.2.5 Myrrh essential oil

This oil contains 32 compounds and the main constitutes are curzerene (40.1%), furanoeudesma-1,3-diene (15.0%), and peleinene (8.4%) [92]. Myrrh essential oil has an anti-inflammatory effect, particularly in acne blemish-prone skin, and is found to be suitable for all types of skin. It provides a healing effect on the facial wound, wrinkle-free facial effects, and reforms the skin elasticity. It also cures the condition of eczema and UV-damaged skin [93–96].

3.1.2.6 Orange blossom essential oil

This essential oil is commonly known as neroli essential oil. Among the 26 components, linalool (29.14%), β-pinene (19.08%), citral (14.1%), limonene (12.04%), *trans*-β-ocimene (6.06%), and *E*-farnesol (5.14%) are found to be the main components [97,98]. It improves the production of sebum and reduces fine lines and flakiness. It also has strong antiseptic properties. Interestingly, the main constituents of this oil boost up the process of new cell regeneration and make the skin scars free, more radiant, and clearer [99–101].

3.1.2.7 Patchouli essential oil
This rejuvenating oil has a strong anti-aging effect. This oil possesses some major components, i.e., Patchouli alcohol (42.75%), delta-Guaiene (28.30%), azulene (20.48%), trans-caryophyllene (11.84%), seychellene (10.77%), naphthalene (8.02%), and cycloheptane (6.02%), respectively [102]. Due to its several beneficial properties, i.e., anti-bacterial, anti-fungal, and antiseptic, it helps in healing the worst facial skin conditions like acne, eczema, dermatitis, and psoriasis [103,104].

3.1.2.8 Rose otto essential oil
Rose otto natural oil comes with lots of medicinal effects such as healing, anti-inflammatory, anti-bacterial, and anti-microbial properties. The β-citronellol (31.15%); trans-geraniol (21.24%), n-heneicosane (9.05%), n-nonadecane (8.77%), nonadecene (4.55%) and phenyl ethyl alcohol (4.16%) are present in most prominent percentages in this oil [105]. This essential oil gives beneficial natural effects in case of eczema, dermatitis, psoriases, and acne blemishes-prone skin. It also revitalizes the skin and reduces UV-induced pigmentation issues [47,106,107].

3.1.2.9 Tea-tree essential oil
Melaleuca alternifolia is known as tea-tree. Due to the high concentration of terpinen-4-ol (48.0%), γ-terpinene (28.0%), 1,8-cineole (15.0%), α-terpinene (13.0%), α-terpineol (8.0%), and *p*-cymene (8.0%), it gives tremendous therapeutic effect in acne vulgaris [108]. It helps in the killing of bacteria, which cause acne such as *Propionibacterium acnes* and hence cure acne scars [109,110]. This oil also maintains the regulation of facial oil and makes the skin acne-free [111].

3.1.2.10 Ylang-ylang essential oil
Ylang-ylang oil is another type of essential oil that is suitable for all types of skin. It possesses some major compounds such as geranyl acetate (18.28%), benzyl benzoate (14.42%), germacrene D (10.92%), trans-caryophyllene

(10.71%), geraniol (8.44%), and eugenol (6.65%) [112]. This oil maintains the overall production of facial oil and skin elasticity. It has anti-inflammatory properties that help in the betterment of inflamed facial skin [113,114].

3.2 Types of facial serums

The choice of facial serum depends merely on the nature of the active ingredient and skin requirements. There are four types of facial serums which are discussed below.

3.2.1 Anti-aging facial serums

These serums contain many anti-aging ingredients individually or in combination, according to the formulation demands. Retinol, vitamin C, vitamin E, ethanolic extract of blackberry (contains a large number of anthocyanins) and methanol plus aqueous-based extract of grape seeds (contains carotenoids and polyphenolic compounds such as proanthocyanidins), growth factor, many peptides (pentapeptides, hexapeptides palmitoyl oligopeptide, palmitoyl tetrapeptide-7), 5-aminolevulinic acid, water-based extract of green tea (main components are epicatechin, epicatechin-3-gallate, epigallocatechin, and epigallocatechin-3-gallate), glycolic acid, alpha hydroxyl acids (AHA), hyaluronic acid, rutin trihydrate, palmitoyl tripeptide, and dipalmitoyl hydroxyproline are the main anti-aging ingredients. These ingredients show a notable difference in skin aging as well as UV-induced facial problems, i.e., improves skin elasticity and tone, fine lines, wrinkles, and sagging problems of the skin [43,115–122]. It has been observed that anti-aging serums give miraculous recovery from skin problems after their application, particularly at night time.

The combination of niacinamide (4%) with kinetin (0.03%) is found to be effective for larger facial pores and gives wrinkle-free skin [123]. Another study has confirmed that the facial sign of aging can be controlled by the use of viper serum. The serum contains a synthetic protein that is derived from *Tropidolaemus wagleri's* venom. Some snake venoms facial serums such as 4% snake active elixir-serum® and *syn*-ake viper serum® are available in the market to rectify the visible sign of aging problems, respectively [124]. The facial application of 0.5% salicin-based serum can improve wrinkles, fine lines, and hyperpigmented marks in 12 weeks [125]. However, the SNAP-8 peptide is also called acetyl glutamyl heptapeptide-1, which is the modified form of Argireline. This peptide has shown an anti-aging and anti-wrinkle effect in 34 women (mild to moderate UV-damaged skin) after 4–12 weeks of C8 Peptide Intensive Treatment® facial serum therapy. According to the above-mentioned clinical trial, the serum works incredibly on the photoaged and damaged skin and reduces the fine lines and crow's feet effect. The main ingredients of the serum were kinetin (growth factor), acetyl octapeptide-3 (SNAP-8 peptide), tocopheryl acetate (vitamin E), tetrahexyldecyl ascorbate (vitamin C), beta-glucan, and sodium hyaluronate. The product is available in the market with a trade name of kinase C8 Peptide Intensive Treatment® [29].

Another marketed anti-aging serum, i.e., Ureadin Fusion Serum Lift Antiarrugas® gives marvelous effect and resolved the skin sagging problem in 28 days in a clinical trial-based study [126]. Several in vitro studies claimed that peptides and growth factors containing topical formulations increase the collagen-induced process and repair the maximum skin problems in a short duration of time [127]. It has been confirmed that the 56 days of post-application of the hyaluronic acid (HA) serum after the single-use of neuromodulator injection significantly reduced the sign of aging [128]. Additionally, L-ascorbic acid, ergothioneine, HA, and fragmented soluble proteoglycan peptides containing facial serum provide maximum protection against the photo-aged skin, minimizes the size of wrinkles, and increases skin hydration within 4 weeks of therapy [23]. However, some clinical data have proved that the application of facial night serum containing ascorbyl tetraisopalmitate, melatonin, and bakuchiol at bedtime plays an important role in skin tightening without any cause of irritation [48,49]. In 56 days of a clinical trial, the palmitoyl pentapeptide-4 (07%) containing facial serum improved the early sign of aging, i.e., periorbital wrinkle with no harmful effect [129]. However, both clinical and in vitro (human fibroblast cells CRL 2097 and non-cell based assay) randomized double-blind studies data confirmed that the facial elasticity has been significantly improved in 56 days by using the 0.5% herbal ethanolic extract-based serum of *Grammatophyllum speciosum* (the active compound is gastrodin) [130]. Some other herbal extracts such as the extract of bakuchiol (Sytenol®, Sytheon Ltd., Rahn France), i.e., natural retinol alternative, *Vanilla tahitensis* (Vanirea®, Solabia Ltd., France) (hydro-glycolic extraction, contain polyphenols including vanillin and parahydroxybenzoic acid), ethanolic extract of *Litchi chinensis* (possess flavonols, i.e., procyanidin B_4, procyanidin B_2, epicatechin, and anthocyanins), and aqueous extract of leaves of *Perilla frutescents* (contain phenolic acids, flavonoids, anthocyanins) based facial serums have been used alone or in combination to slow down the aging process and improve skin elasticity [131–133]. The open-label study of marketed L'Unique Miracular Facial Serum® was found to be safe and it improved the wrinkles and facial fairness after 84 days of usage [44]. Moreover, a split-face controlled randomized clinical trial has confirmed that the use of serum containing vitamin C (20%), tocopherol, and raspberry leaf cell culture extract (contain anthocyanins) gives the anti-aging and skin brightening effect after 60 days of the treatment [134].

3.2.2 Facial brightening and lightening serums

Facial hyperpigmentation issues can be resolved through the use of serums. The hyperpigmentation problems are generally associated with many skin diseases (acne, eczema, perioral dermatitis, rosacea, and psoriasis), drug's side effects (minocycline, antipsychotics, antimalarials, and cytotoxics), UV-radiations, and hypocortisolism called Addison's disease [135–139]. Melasma and hyperpigmentation problems are the main reasons due to which facial brightening and lightening serums are highly in demand.

The skin brightening and lightening preparations may contain ingredients such as strong antioxidants (glutathione, ascorbic acids, and its derivative), niacinamide, alpha, and beta-hydroxy acids, cysteamine, kojic acid, azelaic acid, hydroquinone, *Curcuma longa*, ellagic acid, vitamin E, retinoids, resveratryl triacetate, silymarin, resveratryl triglycolate, ascorbyl tetraisopalmitate and licorice ethanolic extract (contains glabridin) [137,140–147]. According to a double-blind study, 2% hydroquinone gives less depigmentation effect as compared to azelaic acid in 20% quantity [148]. Some investigational studies have also recommended that 2–4% of hydroquinone with the combination of different antioxidants and glycolic acid (GA) gives a better skin lightening effect as compared to that of hydroquinone alone [137]. A couple of retrospective studies have shown that the N-acetyl-4-cysteaminylphenol (4%) gives the best results in melasma [137,149]. The FDA has approved triple action therapy of fluocinolone acetonide (0.01%), hydroquinone (4%), and retinoic acid (0.05%) (tretinoin) for the treatment of melasma [142,150]. It has also been found that the addition of ferulic acid with vitamin E (1%), ascorbic acid (15%), and its derivatives resulted in stable serums and gave double photoprotection against UV-induced hyperpigmentation [151,152]. However, 0.3% of rucinol serum has improved the condition of melasma in 90 days [153]. The clinical trial-based study showed that the naturally occurring polysaccharide, i.e., jellos-based facial serum (oil-in-water) increased the inhibitory action of tyrosinase and reduced the sign of aging as well as improved skin hydration

[154]. A combination of tranexamic acid (3.0%), vitamin B_3 (5.0%), kojic acid (1.0%), and hydroxyethylpiperazineethane sulfonic acid (5.0%) based facial serum was found to be beneficial in hyperpigmentation and dyschromia, i.e., melasma [155]. However, the mixture of some cold press processed natural fruits seeds extract, i.e., pomegranate, blackberry, raspberry, and blueberry (dissolved or suspended in water-lower alcohol mixture) based facial serums give natural skin lightening and glowing effect [156]. Moreover, it has been observed that the combinational use of facial serum containing *Deschampsia antarctica* aqueous extract, vitamin C, and ferulic acid significantly reduced the UV-induced hyperpigmented dark spots very rapidly [31].

An in vitro trial study confirmed that the phytosomes loaded topical serum of *Vitis vinifera* L. powder form seed extract, purchased by Sciyu Biotech Co. Ltd., China (contains epicatechin gallate, procyanidin dimers, tetramers, catechin, epicatechin, gallic acid, procyanidin, resveratrol, flavonoids, and anthocyanins), worked more efficiently with high penetration of the drug [24]. A comparison between the ethanolic extract of cocoa pod husk gel-based serum (that possesses alkaloids, flavonoids, saponins, terpenoids, tannins, polyphenol, monoterpenes, and sesquiterpenes and is cultivated in the region of Jawa Barat, Indonesia) and a marketed Hadalabo ultimate whitening milk serum® has been made in a study. In this study, the phytosomes loaded extract of cocoa pod facial serum was prepared. The resultant complexation of phytosomes with cocoa pod extract was found to be physically and chemically stable. In contrast to marketed serum, the formulated serum showed the same facial skin lightening effect by inhibition of the tyrosine and free radicals [25]. Another study has also been carried out to compare the marketed Neotone® serum with 4.0% hydroquinone test serum (replacement of the Neotone® active with hydroquinone). The result of this study suggested that the formulated serum is more efficient, harmless, and a good alternative in comparison to the marketed serum [157]. It has also been found that the cetyl tranexamate (2%) based facial serum notably decreases the size of hyperpigmented UV-induced patches and inflammation along with face lightening effect [32]. All these lightenings and brightening agents fade the hyperpigmentation marks, reduce the prominent dark spots, and provide more glowing and natural skin.

The results of several studies have also highlighted that many other topical depigmenting and skin lightening agents such as *N*-acetyl-4-S-cysteaminylphenol, glutathione, glycolic acid, azelaic acid (20%), hydroquinone

(4%) and tetrahexyldecyl ascorbate (30%), and retinoic acid (0.1%) may also be used in different skin related problems [15,142,149,158–162].

3.2.3 Anti-acne facial serums

Acne-free skin is possible with the use of anti-acne serums. These serums are very much active and show their effect in a few days of usage. Studies have confirmed that the use of acne-fighting agents in combination with some other ingredients gives more desirable results which include beta-hydroxy acids (BHA), poly-hydroxy acids (PHA), benzoyl peroxide, and vitamin A. They can remove the outermost layer of the skin and control the excess facial oil. They also reduce the appearance of facial redness and acne scars [15,163–169]. Many herbal ethanolic extracts such as *Glycyrrhiza glabra* (contain glabridin) and *Angelica dahurica* (major components are oxypeucedanin hydrate, oxypeucedanin, imperatorin, and isoimperatorin) have shown a notable anti-bacterial activity against acne [170,171].

The clinical study results of purified Apis mellifera (western honey bee) venom facial serum showed a good response in grade 1 and grade 2 acne vulgaris in 42 days. This serum is now available in the market for the treatment/management of acne vulgaris [172,173]. An in vitro study has been conducted to formulate the essential oils-based anti-acne facial serum. The three different types of essential oils containing facial serum, i.e., olive oil, patchouli oil, and lime peel oil (18:01:11) were prepared. The serum showed antibacterial action against the acne-producing gram-positive bacteria, i.e., Cutibacterium acnes (Propionibacterium acnes) [174]. The ethanolic-aqueous extract of pantropical weed, i.e., Euphorbia hirta (asthma plant; contain triterpenoids, coumarins, and diterpenes) was found to be very effective in active acne problems [175]. The application of a marketed Rosa T anti-acne facial serum® (Australian tea tree oil + linoleic acid +vitamin E) after the use of facial cleanser, i.e., Rosa T Mild Cleanser® showed a good antibacterial effect in moderate to severe condition of acne [176]. Some new studies data revealed that nitric oxide generating serum can reduce the size of acne scars and fine lines [177,178].

3.2.4 Hydrating, moisturizing, and facial repairing serums

These serums are usually used for dry flaky and scaly skin. They deliver extra hydration to the deeper layers of the facial skin. The facial repair serums perform a natural facial repairing process with the help of some typical ingredients such as essential oils, amino acids, vitamin B_5, glycerine, and hyaluronic acid [41,42].

A study has been designed to evaluate the anti-wrinkle and hydrating effects of nano-hyaluronic acid-containing topical preparations such as cream, lotion, and serum on 32 women after 2–8 weeks of therapy. According to the results of the study, the preparations helped in the removal of wrinkles and fine lines which appeared due to dehydration of the skin and increased hydration of the skin to almost 96% [179]. The concurrent use of oral zeaxanthin supplements with the facial application of serum containing zeaxanthin, peptides, aqueous extract of algae (mycosporine-like amino acids), and hyaluronic acid can improve facial hydration very rapidly [180]. Similarly, a clinical trial has also been carried out on the use of hyaluronic acid containing serum on the face of 59 women (UV-damaged skin) which significantly improved their skin texture, fine lines, and wrinkles including crow's feet within 4–12 weeks of the treatment [30]. The in situ and ex vivo study of the autologous facial serum loaded with plasma containing growth factors showed strong antioxidant action with a rapid improvement process of skin regeneration [181].

3.3 Hair serums

In the last few years, the popularity and demand for hair serums have increased immensely as these serums help to repair the damaged and dried hair roots and promote healthy hair production. Generally, these serums are liquid-based products. The different types of hair serums, which are widely in use, are discussed below.

3.3.1 Silicone-based serums

These types of serums are more in demand because they mask up the entire hair scalp and protects against humidity, dust, and pollution. They also act as a UV protector and make the hairs strong, shiny, and unbreakable as well as prevent hair loss. In the case of silicone-based serums, the insoluble silicone is stabilized in the form of an emulsion [182–184]. The analysis of the deposition of silicone after the applications of these serums by X-ray fluorescent technique is very useful in the treatment of silicone-based shampoos with different polymers. It has also been suggested that silicone deposition strongly depends on the size and molecular weight of the polymers [185].

A study has been carried out, which demonstrated that the charges of the surface polymers play an essential role in the improvement of the deposition of silicone on hairs. The results showed that an emulsion with a positive surface charge of silicone of amino-modified derivative gives a higher deposition on hairs as compared to that of the polydimethylsiloxane emulsion which had a negative charge [186]. The increased size of different thickening polymer molecules (polyethylene, polyvinyl alcohol, polyisobutylene, and polystyrene) in the serum with the addition of silicone resulted in enhanced silicone deposition on hairs [187,188].

3.3.2 Amino acid-based serums

The amino acid-based serums are used for weak and damaged hair follicles. The hair shaft is generally made up of keratin, and the production of keratin depends on 21 different amino acids (e.g., arginine, leucine, threonine, proline, etc.). Some other natural sources of amino acids also play an essential role in hair growth like red meat, chicken, eggs, fish, potatoes, sesame seeds, soya beans, cabbage, banana, peas, wheat, spinach, rice, corn, and other dairy products. These natural products contain many amino acids such as arginine, cysteine, methionine, lysine, glycine, glutamine, proline, and tyrosine, etc. [189,190].

Arginine, an amino acid in some hair products (conditioners) not only repairs the damaged hair cuticles but also improves the hydrophobicity of the hairs while both phenylalanine and histidine increase the toughness of the hairs. Beta-amino acid and taurine-containing formulations are also beneficial for hair growth. The researchers have shown that 75 mg of taurine improves the quality and density of the hairs in combination with vitamins, polyphenols, lycopene, zinc, omega-3, and omega-6 [191,192]. However, the sulfur-containing methionine amino acid promotes overall hair growth with healthy hair formation [192]. Nitrogen-containing arginine amino acid also enhances the supply of blood to the roots of the hair by unlocking the channels of potassium [193]. It has also been found that L-lysine containing preparations also promote the growth of hair, and therefore, all amino acid-based serums are considered an essential source for the nourishment of hair follicles [194].

3.3.3 Ceramide-based serums

Ceramide lipid is naturally present in the hair cuticles and is also known as hair cuticle cement. This cuticle cement plays a significant role in keeping the pattern of the hair scales properly. However, some studies have proved that the chemical treatment and environmental stress trigger the hair damaging process (dryness, brittleness, hair breakage, and loss of elasticity) by reducing the total quantity of the natural ceramide. Such types of hair serums are beneficial for dry, frizzy, split, and chemical-treated hairs [194–198].

Additionally, many other active ingredients such as the FDA-approved ketoconazole (1–2%)-containing hair products are also used for seborrheic hair dermatitis and fungal infections [199]. On the other hand, hair peptides (palmitoyl pentapeptide-4, copper peptides, and tripeptides) enhance the production of collagen in the hair scalps and encourage the healthy and strong growth of the hairs [118,200,201]. Many vitamins such as biotin, cyanocobalamin, niacin, pantothenic acid, pyridoxal phosphate, thiamine, tocopherol, etc., are also known to improve the hair strength for weak and dull hairs. The pumpkin seed oil can decrease dihydrotestosterone, which is a crucial hormone for hair loss, whereas rosemary gives tensile strength for dry and damaged hairs. Similarly, zinc provides nourishment for dry and frizzy hairs and therefore, has been widely used in different hair products [192]. According to some in vitro studies, caffeine actively works as a healthy hair growth promoter in the case of androgen-dependent hair growth inhibition. The excess quantity of this hormone around the hair follicles tends to reduce the growth of hairs by the lack of hair nutrients supply [192,202,203]. Moreover, the combination of the Saberry®, i.e., extract of amla in powder form (the main components are amino acids such as alanine, aspartic acid, glutamic acid, lysine, and proline) and peanut shell powder (luteolin is the major active constituent) with selenium micronutrient, coconut freeze-dried water, and sandalwood fragrance loaded hair serum has been found to show an excellent effect on hair texture and reduce the hair fall in 90 days of the therapy [204].

3.4 Nail serums

Nail serums are extensively used for weak, scuffed, and brittle nails. These serums nourish the unhealthy nails and cuticles and also restore all the vitamins and minerals. A study has shown that panthenol (2%) accurately recovers the process of nail hydration [205]. The leading cause of brittle and weak nails is the disturbance of the natural process of keratinization due to many diseases like eczema, ichthyosis, psoriasis, and Darier's disease [206,207]. The external application of keratin makes the nail strong and healthy. Antifungal nail serums are also available in the market for the treatment of onychomycosis (tinea unguium). It has been found that ketoconazole improves the condition of onychomycosis by inhibiting the growth of dermatophytes [208].

The vitamin-based nail serums are also used for the improvement of nail cuticles and to treat nail brittleness. The nanoparticles (NPs) of zinc oxide-containing essential oils-based nail serum are available in the market for moisturization of the nails [209]. The results of a double-blind study indicated that 6 months of treatment with the topical application of vitamin E on the affected nail plate (long-standing yellow nail syndrome) enhances the growth of nails [210]. By using the isoamyl lactate as a base in antifungal nail serum, the combination of *Simmondsia chinensis* seed oil (jojoba oil) with the antifungal compounds, i.e., undecylenic acid and synthetic thiocarbamate (tolnaftate) has found to be effective in the treatment of the tinea unguium [211,212]. The data revealed some valuable facts about the significantly increased antifungal activity of the nail serum against the *Trichophyton rubrum* and *Trichophyton mentagrophytes*-induced tinea unguium when thyme essential oil (*Thymus vulgaris*), 1,2-decanediol, polyquaternium-7 and keratolytic agent were used in combination, respectively [213]. Moreover, it has been observed that the cyclodextrin polypseudorotaxanes-based three compounds combination, i.e., methyl sulphonyl methane, biotin, and diethylsilanediol salicylate synergistically improved the condition of onychorrhexis and Beau's lines. This commercial product is available in the market with the trade name of Regenail® [214].

3.5 Other types of serums

Different other types of serums are also available in the market, such as under-eye serums, lip serums, hand serums, and foot serums. Under-eye serums are predominantly used for tightening the fine lines, crow's feet, and wrinkles around the eyes. These serums are also employed to reduce the puffiness below the eye. However, many active agents used in these types of serums have various individual properties such as retinol is used to boost up collagen synthesis and reduces the fine lines and wrinkles, hyaluronic acid improves the skin sagging issues, and peptides are used to rectify the deepest fine lines and wrinkles. Because of their advantages (deep absorption and non-greasy appearance) over the eye formulations like creams, these serums are nowadays used most commonly [215–221].

Lip serums are used for high delivery and deeper penetration of the active agents in dry and scaly lips. These serums are oil-based and may contain vitamin E, olive, shea butter, and avocado oils, and also contain many antioxidants and hyaluronic acid to improve the natural lip color and restore the dryness via deep hydration of the lips skin. The popularity and the use of lip balm over the lip serums have been reduced recently due to its oily nature, low spreadability, and comedogenic, respectively [222].

Additionally, minimally invasive cosmetic procedure, i.e., injectable lip filler is a very popular method for lip augmentation. After reporting some serious disadvantages (lips redness, swelling, bleeding, bruising, and tenderness at the site of injection) and complications (lip asymmetry, the formation of the lumps, allergic reaction, infection, and ulceration at the site of injection), eventually, the demand of the non-invasive cosmetic lip serums treatment has been increased. A study was carried out, which proved that the lips care and protection in a two-phase treatment containing formulations, i.e., lip-renewal (including human growth factors, emollients, tripeptide palmitoyl-glycyl-histidyl-lysine complex, hyaluronic acid, and marine filling spheres) and lip-plumper (including emollients, vitamin B_3, and essential fatty acids), plays an effective and major role in lips moisturizing and augmentation [223].

Hand serums are mostly used to minimize the appearance of wrinkles (anti-aging effect), age spots, and uneven tone. The hand serums usually contain vitamin E, organic shea-oil, organic rice, calendula oil, aqueous extract of lavender, many vitamins, antioxidants, and peptides to restore the elasticity of the skin and hence improve the overall hand appearance [41,224–227]. Moreover, many types of foot serums are also available and are used for dry, scaly, flaky, and damaged foot skin. These serums include various ingredients such as hazelnut oil, walnut oil, apricot kernel oil, avocado oil, and many vitamins (particularly vitamin E) and antioxidants to reduce foot skin infections, wrinkles, dryness, and re-built the foot's flaky skin. Furthermore, the sole serum contains active ingredients such as lidocaine for the pain-relieving purpose which may be caused after wearing high heels, and lavender oil to give anti-bacterial, anti-fungal, and anti-inflammatory effect to make the sole pain and infection-free [228–230]. Recently, nanoparticle-based essential oils containing hand and foot moisturizing serum have been marketed [209].

4. Methods to promote percutaneous absorption of serums

Many cosmetologists are using some valuable procedures for deeper penetration of the serums, which provides rapid action and desired results after a few sessions of the serum treatment. Some of those procedures are discussed as follows.

4.1 Derma roller

The derma roller is the best facial serum penetration approach. The first US patent of this minimally invasive technique was published in 1976 [231]. Recently, this technique has become more popular because it is inexpensive and harmless. Derma roller is an advanced form of micro-needling and contains several micro-needles that puncture the layers of the skin according to their size [232–235]. It has been found that the use of a derma roller stimulates the natural production of collagen and also enhances the percutaneous absorption of the drug molecules. It is used more frequently for the treatment of acne scars, cellulitis, wrinkles, fine lines, aging, and hyperpigmentation issues and also for hair regeneration. The derma roller is also used in all the problems related to hairs and face [232,234–242]. Derma rollers consist of fine needles (0.2–1.5 mm) and have been applied to the skin surface of the head for the activation of the hair follicle [243]. It has also been found that the absorption of minoxidil (5%) solution was increased using this technique in the treatment of alopecia areata [240,244]. In many studies, it has been confirmed that the topical application of platelet-rich plasma (PRP) and minoxidil by derma rollers as well as topical 5% minoxidil with fractional radiofrequency micro-needling improved the condition of alopecia areata and regenerated the hairs from the follicles [243,245,246]. Also, the application of triamcinolone acetonide on the head through a derma roller showed clinically excellent response in alopecia areata [247,248].

Many clinical trials have also confirmed the beneficial role of micro-needling (derma roller) in the treatment of vitiligo [249–252]. Recent clinical studies have shown that the condition of vitiligo improved via the re-pigmentation process after the topical application of 5-fluorouracil solution with derma roller [250,253–257]. Moreover, many other drugs such as topical tacrolimus, calcipotriol with betamethasone, triamcinolone acetonide, and trichloroacetic acid have found to be effective with derma roller in the condition of vitiligo, respectively [258–262].

Furthermore, five facial micro-needling sessions before every 14 days use of amniotic fluid mesenchymal stem cell-derived conditioned media was found to reduce the wrinkles and slow down the aging process [263]. The combinational use of skin lightening serums, i.e., sophora–alpha and rucinol with derma roller gives the most promising result in melasma [237]. The half an hour (twice a week) 8 weeks application of micro-needle patches containing hyaluronic acid, melatonin, vitamin B_3, arbutin, and niacinamide is safe and effective in facial hyperpigmentation condition [264]. Similarly, the micro–needle patches of the mixture of hyaluronic acid, aqueous extract of seaweed (contain polyphenols, alginic acid, and essential amino acid), palmitoyl tripeptide-5, lysine/arginine poly-peptide, acetyl octapeptide-3, and adenosine gives a beneficial effect on wrinkled skin and improves facial skin dryness [265]. Additionally, the single application of 4% hydroquinone serum with derma roller before the 1-month treatment of modified Kligman's formula, i.e., fluocinolone acetonide 0.01%, hydroquinone 4.0%, and tretinoin 0.05% along with sun protective factor 70 (SPF-70) improved the condition of melasma very rapidly [266]. The results of many clinical trials suggested that tranexamic acid, protein-rich plasma, and L-ascorbic acid work efficiently with derma roller and swiftly recover the melasma condition [267–271]. Different types of derma rollers are shown in Figs. 3 and 4 and different sizes of the rollers and their indications are given in Table 2. However, the use of this smart approach depends on the skin thickness, need, and demand of the diseases.

4.1.1 Contraindication
Derma rollers are effectively used in all facial and hair problems. These derma rollers also produce grievous effects and are contraindicated in conditions like acne vulgaris, eczema, psoriasis, rosacea, herpes labialis, and skin malignancy. Patients taking anticoagulants are also contraindicated for the use of these rollers [244].

4.2 Iontophoresis
The iontophoresis method is widely used in the field of dermatology and cosmetology. The word iontophoresis comes from the Greek word

Fig. 3 Different sizes of needles of derma roller.

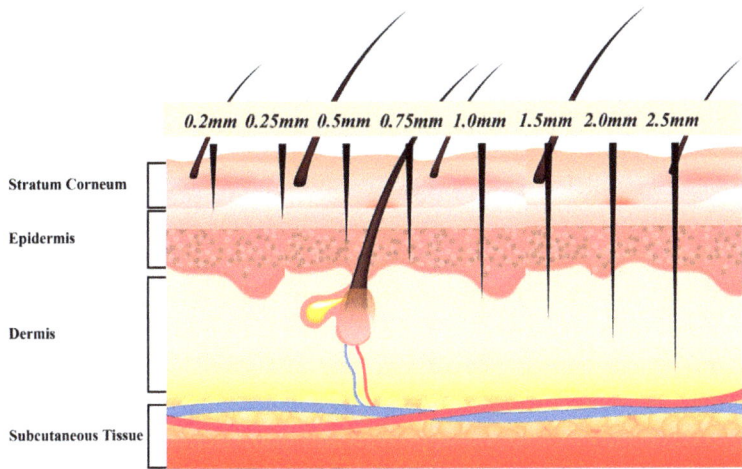

Fig. 4 Depth of different needles into the skin.

Table 2 Sizes of derma roller needles with indications.

Indications	Needle sizes (mm)							
	0.2	0.25	0.5	0.75	1.0	1.5	2.0	2.5
Hair regrowth and regeneration	✓	✓	✓	✓	✓	✓	✗	✗
Fine lines	✗	✓	✓	✗	✗	✗	✗	✗
Anti-aging and anti-wrinkle	✗	✗	✓	✓	✓	✗	✗	✗
Open pores	✗	✗	✓	✓	✓	✗	✗	✗
Pigmentation dark spots	✗	✗	✓	✓	✓	✗	✗	✗
Acne scars	✗	✗	✓	✓	✓	✗	✗	✗
Light color scars	✗	✗	✓	✓	✓	✗	✗	✗
Cellulitis	✗	✗	✗	✗	✓	✓	✓	✗
Scratch marks	✗	✗	✗	✗	✓	✓	✓	✓
Deep dark scars	✗	✗	✗	✗	✗	✓	✓	✓

"ionto" which means "ion" and "phoresis" means "carried away", respectively. This method increases the topical drug delivery to the skin by electromotive force and delivers the molecules of the drug beneath the skin [272–275]. The iontophoresis device is called the galvanic iontophoresis. In order to resolve the facial skin problems, FDA has approved two types of portable iontophoresis devices that have been available in the market and used extensively for the correction of facial wrinkles (Wrinkle-MD® for the deep penetration of peptides and hyaluronic acid) and face lightening as well as brightening purposes (NuFACE® Trinity) [276]. These devices produce the electrical current and communicate with the first and second electrodes. These electrodes ionically communicate with the human skin, which results in the passage of electrons between these two conductive electrodes with the standard potential difference of at least 0.2 V. The charged and hydrophilic drug molecules give the most promising results via iontophoresis. However, $\leq 0.5 \, mA/cm^2$ is a physiologically approved and harmless current for human skin [248,274–283].

The value of this nonsurgical/noninvasive drug delivery system has been enhanced recently due to the improved and good penetration of lipophilic, hydrophilic, and high molecular weight drug molecules in 10–30 min without any facial burning and irritation [279,280,284]. However, the incorporation of different drug penetration enhancers (dimethyl sulfoxide, oxazolidinone, dimethylformamide, fatty acids, and terpenes, etc.) into any topical formulation increases the two folds transfer of the active ingredient(s) beneath the stratum corneum. The application of such topical formulations has been ameliorated more quickly and efficiently through the iontophoresis technique [275,285–287].

A study was designed to investigate the effect of vitamin C on the condition of melasma using the iontophoresis technique. According to the results of the study, vitamin C was found to be a potent inhibitor of melanogenesis, and the melasma-containing area gradually reduced after 6 weeks of treatment (two times/week for 15 min) [288]. According to the findings of various clinical trials, the non-invasive vitamin C iontophoresis technique has been used more efficiently and successfully for the treatment of melasma, rolling scars as well as post-acne vulgaris hyperpigmentation marks, respectively [289–293]. The results of another comparative study of glycolic acid (70%) peel and topical vitamin C nanosome iontophoresis also suggested that the topical vitamin C technique is more effective, harmless, reliable, and non-invasive as compared to glycolic acid peel [294]. However, the use of mandelic acid serum and broad-spectrum SPF after the vitamin C iontophoresis treatment gives long-term management in melasma [295]. Another in vitro and in vivo comparative study showed that 12 weeks of treatment of multivitamins through iontophoresis exhibited an excellent comparative depigmentation effect than vitamin C alone [296]. Similarly, some spilled-face study results also indicated that the cutaneous absorption of L-ascorbic acid (23.8%) was enhanced when used with iontophoresis. It also significantly reduced the UV-induced skin aging after the cessation of treatment within 2 weeks [34].

The iontophoretic devices are also used for hair growth stimulation and ionic drug absorption. These small devices contain electrodes (cathode, anode), which penetrate the drug rapidly and easily without any harm to the patients [297]. A study was carried out to observe the effect of iontophoretic treatment of topical minoxidil sulfate. The results indicated that it is beneficial for the treatment of alopecia and provides significant penetration of the drug through the hair follicles [298]. Moreover, it also increases the transungual permeation of the antifungal drug delivery in the case of nail fungal infections such as onychomycosis [207,299]. In the comparison of oral therapy, topical drug therapy gives a more targeted effect and bypasses the systemic absorption associated side effects in onychomycosis condition. The iontophoresis technique has been used to avoid inadequate transungual drug absorption [300–302]. However, the transungual application of griseofulvin through iontophoresis increased its antifungal action by approximately eight folds [303,304]. Also, the penetration of the antifungal drug across the corpus unguis (nail plate) was enhanced via enhancers such as carbocysteine, hydroxypropyl-b-cyclodextrin, N-acetylcysteine, thioglycolic acid, and sodium lauryl sulfate, etc., and their combinational use with iontophoresis gave a more significant and reliable effect than the iontophoresis alone [302,305–308]. A comparative clinical trial study was carried out for triamcinolone acetonide with iontophoresis once a month (for 6 months) and daily corpus unguis psoriasis application of calcipotriol with betamethasone dipropionate, respectively. According to the results of the trial, it was found that the monthly application of triamcinolone acetonide with iontophoresis for 20 min improved the therapeutic effect with an advantage of deep penetration as similar to the daily topical application of calcipotriol with betamethasone dipropionate [309]. A recent study has proved that the use of iontophoresis on adhesive micro-needle skin patches of hyaluronic acid remarkably improved the delivery of the drug in a short duration of time [310].

4.2.1 Limitations

The iontophoresis technique is free from side effects. Despite the fact, minor skin reactions (minor irritation, tingling sensation, itching, and erythema) have been reported due to the increased current exposure time and intensity. Nevertheless, it is contraindicated on deep flaky, broken, inflamed, and ruptured skin surface [276,311,312].

4.3 Phonophoresis or sonophoresis

The phonophoresis/sonophoresis is a safe, harmless, and non-surgical-based ultrasound wave method. In this method, the active compound molecules penetrate the deepest layers of the skin through ultrasound waves. The low (20–100 kHz) and medium (0.2–2 MHz) frequencies of ultrasound waves are used according to the drug molecules. The low-frequency phonophoresis (LFP) is used for the deep penetration of high molecular weight drugs or compounds such as nanoparticles, hormones, ascorbic acid, proteins, etc. The medium frequency phonophoresis (MFP) is used for the delivery of low molecular weight drugs such as topical steroids and non-steroidal anti-inflammatory drugs (NSAIDs). However, the high frequency of the ultrasound waves (2–10 MHz) is employed for physical therapy (PT), imaging, and the pulverization of the kidney and gall stones, respectively [248,313–317].

The clinical trial in humans showed that the 10 min treatment of both, i.e., cyclosporine solution and methyl-prednisolone ointment via LFP (25 kHz) after every 21 days for 3 months incredibly reduces the spot baldness as well as boost the growth of hair [318]. Moreover, the facial application of vitamin C (20%) with 2% gel of kojic acid via phonophoresis/sonophoresis method was found to improve the condition of facial melasma in 35 days [319]. In the case of chloasma, it has been observed that the facial application of arbutin B_3 and levorotatory L-ascorbic acid simultaneously through the phonophoresis/sonophoresis method provided approximately 98% improvement in 42 days of the clinical study [320]. The antioxidant compound LycogenTM is used for facial skin rejuvenation. The clinical trial study (10 weeks) has been conducted on the face of 26 females (between 30 and 45 years old). The result of the trial showed that the facial application of LycogenTM via phonophoresis/sonophoresis technique act as a most promising and appropriate antioxidant compound in the case of hyperpigmentation, poor facial skin texture, and wrinkles [321]. The application of vitamin A serum (0.3–0.5%) on the face by phonophoresis/sonophoresis was found to effective in skin dryness and reduced facial hyperpigmentation as well as produced anti-aging and skin lightening effects [322].

4.3.1 Limitations

Besides the fact, it is a long and time-consuming technique, sometimes, irritation and burning sensation with tingling effects may be produced on the skin of sensitive people [313].

4.4 Gemstone roller

Gemstone facial massager roller is one of the best traditional Chinese beauty skincare tools, made up of green/white jade, rose quartz, and black obsidian gemstones. Lately, the use of gemstone facial massager roller has been increased but it is not a new invention. Chinese women have employed gemstone rollers as skincare applicators since the 12th century because they believed that these gemstone rollers reduce puffiness, improves the circulation of blood, and reduce facial wrinkles as well as fine lines [323,324].

The facial massager roller is composed of different gemstones, connecting by a rod as a supporter. In the case of a double-face massager roller, both ends of the joining rod are screwed to the fixing bolt cap and the shaft sleeve is connected with the shaft (supporter). The large gemstone (for the overall face) is connected at one end and the small gemstone (for under-eye) is connected to another end of the rod. Moreover, the battery is installed in the joining rod with the charging site and on/off button, respectively. The inbuilt vibrating motor boost up the facial blood circulation, improve facial skin elasticity, eliminate toxin, improve facial tone, encourage the lymphatic drainage, reduce facial puffiness as well as dark under-eye circles. After the application of the beauty product(s), i.e., facial serum, lotion, and cream, the use of a gemstone facial roller provides deep penetration and gives incredibly more youthful facial skin [323,324].

5. Conclusion

On account of several beneficial effects in a short duration of time with relatively low or no side effects, the attraction of people toward the pharmaceutical-based cosmetic serums have been ameliorated very rapidly. The popularity and demand of these cosmetic serums have been enhanced quickly globally. A vast range of different types of cosmetic serums are available and used according to skin types and problems. However, various pure essential oils, herbs, vitamins, peptides, strong antioxidants, and chemical exfoliates-loaded serums give a rapid and targeted effect instead of many other types of marketed cosmetic products (creams, gels, foams, and lotions, etc.). According to many different studies, the application of cosmetic serums with the help of iontophoresis, phonophoresis/sonophoresis, derma roller, or gemstone roller methods give more consistent and deep penetration over a prolonged period. Consequently, the demand for other cosmetic products is badly affected by cosmetic serums. That day is now not very far when pharmaceutical-based cosmetic serums will replace other types of cosmetic dosage forms in the future.

References

[1] S.V. Sastry, J.R. Nyshadham, J.A. Fix, Recent technological advances in oral drug delivery—a review, Pharm. Sci. Technol. Today 3 (2000) 138–145.

[2] J. Roy, Pharmaceutical impurities—a mini-review, AAPS PharmSciTech 3 (2002) 1–8.

[3] Z. Ayenew, V. Puri, L. Kumar, A.K. Bansal, Trends in pharmaceutical taste masking technologies: a patent review, Recent Pat. Drug Deliv. Formul. 3 (2009) 26–39.

[4] L.L. Augsburger, S.W. Hoag (Eds.), Pharmaceutical Dosage Forms—Tablets, third ed., CRC Press, Boca Raton, FL, 2016.

[5] L. Allen, H.C. Ansel, Ansel's Pharmaceutical Dosage Forms and Drug Delivery Systems, nineth ed., Lippincott Williams & Wilkins, Philadelphia, USA, 2013, pp. 90–143.

[6] M. Gibson (Ed.), Pharmaceutical Preformulation and Formulation: A Practical Guide From Candidate Drug Selection to Commercial Dosage Form, second ed., CRC Press, Florida, USA, 2016, pp. 157–223.

[7] L.T. Smith, K.A. Holbrook, Embryogenesis of the dermis in human skin, Pediatr. Dermatol. 3 (1986) 271e80.

[8] J. Kanitakis, Anatomy, histology and immunohistochemistry of normal human skin, Eur. J. Dermatol. 12 (2002) 390–401.

[9] P.A. Kolarsick, M.A. Kolarsick, C. Goodwin, Anatomy and physiology of the skin, J. Dermatol. Nurses Assoc. 3 (2011) 203–213.

[10] K. Kabashima, T. Honda, F. Ginhoux, G. Egawa, The immunological anatomy of the skin, Nat. Rev. Immunol. 19 (2019) 19–30.

[11] L. Fodor, D. Dumitrascu, Skin Anatomy, in: L. Fodor, Y. Ullmann (Eds.), Aesthetic Applications of Intense Pulsed Light, Springer, Cham, Switzerland, 2020, pp. 1–12.

[12] J. Schauber, R.L. Gallo, Antimicrobial peptides and the skin immune defense system, J. Allergy Clin. Immunol. 122 (2008) 261–266.

[13] M.L. Bennett, R.L. Henderson Jr., Introduction to cosmetic dermatology, Curr. Probl. Dermatol. 15 (2003) 43–83.

[14] U. Wollina, A. Goldman, U. Berger, M.B. Abdel-Naser, Esthetic and cosmetic dermatology, Dermatol. Ther. 21 (2008) 118–130.

[15] K.A. Bellad, B.K. Nanjwade, M.S. Kamble, T. Srichana, N.F. Idris, Development of cosmeceuticals, World J. Pharm. Pharm. Sci. 6 (2017) 643–691.

[16] American Society for Dermatologic Surgery (ASDS), ASDS Members Performed More Than 12.5 Million Treatments in 2018. Survey Data, Available at: www.asds. net. (accessed 02.01.21).

[17] H.N. Yusoh, N.H. Jong, N.M. Isa, N. Sulaiman, W.H. Yusof, Enhancing vitamin C content in phyllanthus emblica facial serum through cold pressed method, Int. J. Eng. Adv. Res. 1 (2020) 6–16.

[18] S. Budiasih, I. Masyitah, K. Jiyauddin, M. Kaleemullah, A.D. Samer, A.M. Fadli, Y. Eddy, Formulation and characterization of cosmetic serum containing argan oil as moisturizing agent, in: BROMO Conference Proceedings, Shah Alam, Selangor Darul Ehsan, Malaysia, 2018, pp. 297–304.

[19] A. Pandey, G.K. Jatana, S. Sonthalia, Cosmeceuticals, in: StatPearls [Internet], StatPearls Publishing, Treasure Island, FL, 2021.

[20] FDA (Food and Drug Administration), Cosmeceutical, Available at: https://www.fda.gov/cosmetics/cosmetics-labeling-claims/cosmeceutical. (accessed on April 2021).

[21] V.P. Shah, F.Ş. Rădulescu, D.S. Miron, A. Yacobi, Commonality between BCS and TCS, Int. J. Pharm. 509 (2016) 35–40.

[22] S. Sasidharan, J.P. Junise, Formulation and evaluation of fairness serum using poly-herbal extracts, Int. J. Pharm. 4 (2014) 105–112.

[23] A. Garre, M. Narda, P. Valderas-Martinez, J. Piquero, C. Granger, Antiaging effects of a novel facial serum containing L-ascorbic acid, proteoglycans, and proteoglycan-stimulating tripeptide: ex vivo skin explant studies and in vivo clinical studies in women, Clin. Cosmet. Investig. Dermatol. 11 (2018) 253–263.

[24] S. Surini, H. Mubarak, D. Ramadon, Cosmetic serum containing grape (Vitis vinifera L.) seed extract phytosome: formulation and in vitro penetration study, J. Young Pharm. 10 (2018) S51–S55.

[25] S.E. Priani, S. Aprilia, R. Aryani, L. Purwanti, Antioxidant and tyrosinase inhibitory activity of face serum containing cocoa pod husk phytosome (Theobroma cacao L.), J. Appl. Pharm. Sci. 9 (2019) 110–115.

[26] N.H. Cox, Dermatology—just the facts, Br. J. Dermatol. 150 (2004) 155–177.

[27] E.D. Monteiro, L.S. Baumann, The science of cosmeceuticals, Expert Rev. Dermatol. 1 (2006) 379–389.

[28] L. Baumann, Skin ageing and its treatment, J. Pathol. 211 (2007) 241–251.

[29] F. Mccall-Perez, T.J. Stephens, J.H. Herndon Jr., Efficacy and tolerability of a facial serum for fine lines, wrinkles, and photodamaged skin, J. Clin. Aesthet. Dermatol. 4 (2011) 51–54.

[30] S. Raab, M. Yatskayer, S. Lynch, M. Manco, C. Oresajo, Clinical evaluation of a multi-modal facial serum that addresses hyaluronic acid levels in skin, J. Drugs Dermatol. 16 (2017) 884–890.

[31] M. Milani, B. Hashtroody, M. Piacentini, L. Celleno, Skin protective effects of an anti-pollution, antioxidant serum containing Deschampsia antartica extract, ferulic acid and vitamin C: a controlled single-blind, prospective trial in women living in urbanized, high air pollution area, Clin. Cosmet. Investig. Dermatol. 12 (2019) 393–399.

[32] I.D. da Silva Souza, L. Lampe, D. Winn, New topical tranexamic acid derivative for the improvement of hyperpigmentation and inflammation in the sun-damaged skin, J. Cosmet. Dermatol. 19 (2020) 1–5.

[33] E. Markiewicz, O.C. Idowu, Melanogenic difference consideration in ethnic skin type: a balance approach between skin brightening applications and beneficial sun exposure, Clin. Cosmet. Investig. Dermatol. 13 (2020) 215–232.

[34] T.-H. Xu, J.Z.S. Chen, Y.-H. Li, Y. Wu, Y.-J. Luo, X.-H. Gao, H.-D. Chen, Split-face study of topical 23.8% L-ascorbic acid serum in treating photo-aged skin, J. Drugs Dermatol. 11 (2012) 51–56.

[35] Y.S. Kim, Dropper-type cosmetics container in which different types of contents can be used in mixed manner, United States Patent 9,351,556, 2016.

[36] B.A. Ganter, Leak resistant droppers, The United States Patent 16/581,722, 2020.

[37] K. Eshun, Q. He, Aloe vera: a valuable ingredient for the food, pharmaceutical and cosmetic industries—a review, Crit. Rev. Food Sci. Nutr. 44 (2004) 91–96.

[38] J. Hamman, Composition and applications of aloe vera leaf gel, Molecules 13 (2008) 1599–1616.

[39] K. Manvitha, B. Bidya, Aloe vera: a wonder plant its history, cultivation and medicinal uses, J. Pharmacogn. Phytochem. 2 (2014) 85–88.

[40] S. Ojha, S. Sinha Sinha, S.D. Chaudhuri, H. Chadha, B. Aggarwal, S.M. Jain, A. Meenu, Formulation and evaluation of face serum containing bee venom and aloe vera gel, World J. Pharm. Res. 8 (2019) 1100–1105.

[41] C.M. Choi, D.S. Berson, Cosmeceuticals, Semin. Cutan. Med. Surg. 25 (2006) 163–168.

[42] W.P. Werschler, N.S. Trookman, R.L. Rizer, E.T. Ho, R. Mehta, Enhanced efficacy of a facial hydrating serum in subjects with normal or self-perceived dry skin, J. Clin. Aesthet. Dermatol. 4 (2011) 51–55.

[43] N. Khan, S. Ahmed, M.A. Sheraz, Z. Anwar, I.A. Siddique, I. Ahmad, Formulation and stability of anti-aging serum containing rutin trihydrate, Pakistan Patent Application No. 218/2021, 2021.

[44] G. Sadowski, J. Sadowski, Safety and efficacy of a novel antiaging skin care regimen containing neutraceuticals and growth factors on the facial skin of women: a 12-week open-label study, J. Clin. Aesthet. Dermatol. 13 (2020) 24–34.

[45] B.J. West, R. Sabin, Efficacy of a *Morinda citrifolia* based skin care regimen, Curr. Res. J. Biol. Sci. 5 (4) (2012) 310–314.

[46] T.K. Lin, L. Zhong, J. Santiago, Anti-inflammatory and skin barrier repair effects of topical application of some plant oils, Int. J. Mol. Sci. 19 (2018) 70.

[47] K.F. Wolfe, Facial hydrating oil, The United States Patent 16/424,695, 2019.

[48] D.J. Goldberg, D. Mraz-Robinson, C. Granger, Efficacy and safety of a 3-in-1 anti-aging night facial serum containing melatonin, bakuchiol, and ascorbyl tetraisopalmitate through clinical and histological analysis, J. Cosmet. Dermatol. 19 (2020) 884–890.

[49] D.J. Goldberg, D.M. Robinson, C. Granger, Clinical evidence of the efficacy and safety of a new 3-in-1 anti-aging topical night serum-in-oil containing melatonin, bakuchiol, and ascorbyl tetraisopalmitate: 103 females treated from 28 to 84 days, J. Cosmet. Dermatol. 18 (2019) 806–814.

[50] N. Pazyar, R. Yaghoobi, M.R. Ghassemi, A. Kazerouni, E. Rafeie, N. Jamshydian, Jojoba in dermatology: a succinct review, G. Ital. Dermatol. Venereol. 148 (2013) 687–691.

[51] D. Guillaume, D. Pioch, Z. Charrouf, Argan [Argania spinosa (L.) skeels] oil, in: M. Ramadan (Ed.), Fruit Oils: Chemistry and Functionality, Springer, Cham, Switzerland, 2019, pp. 317–352.

[52] A. Orchard, S. Van Vuuren, Commercial essential oils as potential antimicrobials to treat skin diseases, Evid. Based Complement. Alternat. Med. 2017 (2017) 4517971.

[53] S.I. Kirbaslar, F.G. Kirbaslar, U. Dramur, Volatile constituents of Turkish bergamot oil, J. Essent. Oil Res. 12 (2000) 216–220.

[54] M. Nikolić, D. Stojković, J. Glamočlija, A. Ćirić, T. Marković, M. Smiljković, M. Soković, Could essential oils of green and black pepper be used as food preservatives? J. Food Sci. Technol. 52 (2015) 6565–6573.

[55] S. Amiri, S. Sharafzadeh, Essential oil components of German chamomile cultivated in Firoozabad, Iran. Orient. J. Chem. 30 (2014) 365–367.

[56] S.A. Selim, M.E. Adam, S.M. Hassan, A.R. Albalawi, Chemical composition, antimicrobial and antibiofilm activity of the essential oil and methanol extract of the Mediterranean cypress (Cupressus sempervirens L.), BMC Complement. Altern. Med. 14 (2014) 1–8.

[57] W. Deng, K. Liu, S. Cao, J. Sun, B. Zhong, J. Chun, Chemical composition, antimicrobial, antioxidant, and antiproliferative properties of grapefruit essential oil prepared by molecular distillation, Molecules 25 (2020) 217.

[58] P.K. Sharma, V. Singh, M. Ali, Chemical composition and antimicrobial activity of fresh rhizome essential oil of Zingiber officinale roscoe, Pharm. J. 8 (2016) 185–190.

[59] E. Schmidt, S. Bail, G. Buchbauer, I. Stoilova, A. Krastanov, A. Stoyanova, L. Jirovetz, Chemical composition, olfactory evaluation and antioxidant effects of the essential oil of Origanum majorana L. from Albania, Nat. Prod. Commun. 3 (2008) 1051–1056.

[60] D. Lara, E. Vilcacundo, C. Carrillo, C. Carpio, M. Silva, M. Alvarez, W. Carrillo, Obtention of protein concentrate and polyphenols from macadamia (Macadamia integrifolia) with aqueous extraction method, Asian J. Pharm. Clin. Res. 10 (2017) 138–142.

[61] M. Viuda-Martos, Y. Ruiz-Navajas, J. Fernández-López, J.A. Pérez-Álvarez, Chemical composition of mandarin (C. reticulata L.), grapefruit (C. paradisi L.), lemon (C. limon L.) and orange (C. sinensis L.) essential oils, J. Essent. Oil Bear Plants 12 (2009) 236–243.

[62] W.C. Zeng, Z. Zhang, H. Gao, L.R. Jia, Q. He, Chemical composition, antioxidant, and antimicrobial activities of essential oil from pine needle (Cedrus deodara), J. Food Sci. 77 (2012) C824–C829.

[63] K.P. Padmakumari, I. Sasidharan, M.M. Sreekumar, Composition and antioxidant activity of essential oil of pimento (Pimenta dioica (L) Merr.) from Jamaica, Nat. Prod. Res. 25 (2011) 152–160.

[64] O. Boussaada, R. Chemli, Chemical composition of essential oils from flowers, leaves and peel of *Citrus aurantium* L. var. amara from Tunisia, J. Essent. Oil Bear Plants 9 (2006) 133–139.

[65] M.M. Özcan, J.C. Chalchat, Chemical composition and antifungal activity of rosemary (Rosmarinus officinalis L.) oil from Turkey, Int. J. Food Sci. Nutr. 59 (2008) 691–698.

[66] L.D. Sampaio, J.G. Maia, A.M. de Parijós, R.Z. de Souza, L.E. Barata, Linalool from rosewood (Aniba rosaeodora Ducke) oil inhibits adenylate cyclase in the retina, contributing to understanding its biological activity, Phytother. Res. 26 (2012) 73–77.

[67] A.I. Hussain, F. Anwar, M. Shahid, M. Ashraf, R. Przybylski, Chemical composition, and antioxidant and antimicrobial activities of essential oil of spearmint (Mentha spicata L.) from Pakistan, J. Essent. Oil Res. 22 (2010) 78–84.

[68] M.M. Özcan, L.G. Pedro, A.C. Figueiredo, J.G. Barroso, Constituents of the essential oil of sea fennel (Crithmum maritimum L.) growing wild in Turkey, J. Med. Food 9 (2006) 128–130.

[69] J. Disket, S. Mann, R.K. Gupta, A review on spikenard (Nardostachys jatamansi DC.)-an 'endangered' essential herb of India, Int. J. Pharm. Chem. 2 (2012) 52–60.

[70] A. Porte, R.L. Godoy, L.H. Maia-Porte, Chemical composition of sage (Salvia officinalis L.) essential oil from the Rio de Janeiro state (Brazil), Rev Bras de Plantas Medicinais 15 (2013) 438–441.

[71] B. Luo, E. Kastrat, T. Morcol, H. Cheng, E. Kennelly, C. Long, Gaultheria longibracteolata, an alternative source of wintergreen oil, Food Chem. 342 (2021) 128244.

[72] S. Singh, A. Lohani, A.K. Mishra, A. Verma, Formulation and evaluation of carrot seed oil-based cosmetic emulsions, J. Cosmet. Laser Ther. 21 (2019) 99–107.

[73] M. Staniszewska, J. Kula, M. Wieczorkiewicz, D. Kusewicz, Essential oils of wild and cultivated carrots—the chemical composition and antimicrobial activity, J. Essent. Oil Res. 17 (2005) 579–583.

[74] M.M. Özcan, J.C. Chalchat, Chemical composition of carrot seeds (*Daucus carota* L.) cultivated in Turkey: characterization of the seed oil and essential oil, Grasas Aceites 58 (2007) 359–365.

[75] S.F. Van Vuuren, G.P. Kamatou, A.M. Viljoen, Volatile composition and antimicrobial activity of twenty commercial frankincense essential oil samples, South Afr. J. Bot. 76 (2010) 686–691.

[76] A.R. Al-Yasiry, B. Kiczorowska, Frankincense-therapeutic properties, Adv. Hyg. Exp. Med. 70 (2016) 380–391.

[77] R. Karthikeyan, G. Kanimozhi, N.R. Madahavan, B. Agilan, M. Ganesan, N.R. Prasad, P. Rathinaraj, Alpha-pinene attenuates UVA-induced photoaging through inhibition of matrix metalloproteinases expression in mouse skin, Life Sci. 217 (2019) 110–118.

[78] E. Hwang, H.T. Ngo, B. Park, S.A. Seo, J.E. Yang, T.H. Yi, Myrcene, an aromatic volatile compound, ameliorates human skin extrinsic aging via regulation of MMPs production, Am. J. Chin. Med. 45 (2017) 1113–11124.

[79] B. Salehi, S. Upadhyay, I. Erdogan Orhan, A. Kumar Jugran, L.D.S. Jayaweera, D.A. Dias, F. Sharopov, Y. Taheri, N. Martins, N. Baghalpour, W.C. Cho, Therapeutic potential of α-and β-pinene: a miracle gift of nature, Biomolecules 9 (2019) 738.

[80] F.S. Sharopov, H. Zhang, W.N. Setzer, Composition of geranium (Pelargonium graveolens) essential oil from Tajikistan, Am. J. Essent. Oils Nat. Prod. 2 (2014) 13–16.

[81] F.D. Pereira, J.M. Mendes, I.O. Lima, K.S. Mota, W.A. Oliveira, E.D. Lima, Antifungal activity of geraniol and citronellol, two monoterpenes alcohols, against Trichophyton rubrum involves inhibition of ergosterol biosynthesis, Pharm. Biol. 53 (2015) 228–234.

[82] M.C. Leite, A.P. de Brito Bezerra, J.P. de Sousa, L.E. de Oliveira, Investigating the antifungal activity and mechanism (s) of geraniol against Candida albicans strains, Med. Mycol. J. 53 (2015) 275–284.

[83] R. Gaonkar, P.K. Avti, G. Hegde, Differential antifungal efficiency of geraniol and citral, Nat. Prod. Commun. 13 (2018) 1609–1614.

[84] W. Chen, A.M. Viljoen, Geraniol—a review of a commercially important fragrance material, South Afr. J. Bot. 76 (2010) 643–651.

[85] B.R. Rao, P.N. Kaul, K.V. Syamasundar, S. Ramesh, Water soluble fractions of rose-scented geranium (Pelargonium species) essential oil, Bioresour. Technol. 84 (2002) 243–246.

[86] S. Abe, N. Maruyama, K. Hayama, S. Inouye, H. Oshima, H. Yamaguchi, Suppression of neutrophil recruitment in mice by geranium essential oil, Mediat. Inflamm. 13 (2004) 21–24.

[87] W. Mączka, K. Wińska, M. Grabarczyk, One hundred faces of geraniol, Molecules 25 (2020) 3303.

[88] K. Smigielski, A. Raj, K. Krosowiak, R. Gruska, Chemical composition of the essential oil of Lavandula angustifolia cultivated in Poland, J. Essent. Oil Bear Plants 12 (2009) 338–347.

[89] H.M. Cavanagh, J.M. Wilkinson, Biological activities of lavender essential oil, Phytother. Res. 16 (2002) 301–308.

[90] H.M. Cavanagh, J.M. Wilkinson, Lavender essential oil: a review, Aust Infec Cont. 10 (2005) 35–37.

[91] M. Lis-Balchin (Ed.), Lavender: The Genus Lavandula, Taylor & Francis, New York, USA, 2003.

[92] K. Morteza-Semnani, M. Saeedi, Constituents of the essential oil of Commiphora myrrha (Nees), Engl. Var. Molmol. J. Essent. Oil Res. 15 (2003) 50–51.

[93] E.S. El Ashry, N. Rashed, O.M. Salama, A. Saleh, Components, therapeutic value and uses of myrrh, Pharmazie 58 (2003) 163–168.

[94] M. Lemenith, D. Teketay, Frankincense and myrrh resources of Ethiopia: II. Medicinal and industrial uses, SINET Ethiopian J. Sci. 26 (2003) 161–172.

[95] K. Laird, C. Phillips, Vapour phase: a potential future use for essential oils as antimicrobials? Lett. Appl. Microbiol. 54 (2012) 169–174.

[96] N.M. Al-Ghaban, N.B. Kamil, Evaluation of effect of local exogenous application of Myrrh oil on healing of wound incisions of facial skin (histochemical, histological and histomorphometrical study in rabbits), J. Baghdad Coll. Dent. 31 (2019) 71–78.

[97] E. Sarrou, P. Chatzopoulou, K. Dimassi-Theriou, I. Therios, Volatile constituents and antioxidant activity of peel, flowers and leaf oils of Citrus aurantium L. growing in Greece, Molecules 18 (2013) 10639–10647.

[98] C. Federman, C. Ma, D. Biswas, Major components of orange oil inhibit Staphylococcus aureus growth and biofilm formation, and alter its virulence factors, J. Med. Microbiol. 65 (2016) 688–695.

[99] V. Jeannot, J. Chahboun, D. Russell, P. Baret, Quantification and determination of chemical composition of the essential oil extracted from natural orange blossom water (Citrus aurantium L. ssp. Aurantium), Int. J. Aromather. 15 (2005) 94–97.

[100] K. Fisher, C. Phillips, Potential antimicrobial uses of essential oils in food: is citrus the answer? Trends Food Sci. Technol. 19 (2008) 156–164.

[101] M. Vigan, Essential oils: renewal of interest and toxicity, Eur. J. Dermatol. 20 (2010) 685–692.

[102] D. Ermaya, S.P. Sari, A. Patria, F. Hidayat, F. Razi, Identification of patchouli oil chemical components as the results on distillation using GC-MS, in: IOP Conference Series: Earth and Environmental Science, Banda Aceh, Indonesia, 2019.

[103] A. Donelian, L.H. Carlson, T.J. Lopes, R.A. Machado, Comparison of extraction of patchouli (Pogostemon cablin) essential oil with supercritical CO2 and by steam distillation, J. Supercrit. Fluids 48 (2009) 15–20.

[104] P. Tongnuanchan, S. Benjakul, Essential oils: extraction, bioactivities, and their uses for food preservation, J. Food Sci. 79 (2014) R1231–R1249.

[105] T. Atanasova, M. Kakalova, L. Stefanof, M. Petkova, A. Stoyanova, S. Damyanova, M. Desyk, Chemical composition of essential oil from Rosa Damascena mill., growing in new region of Bulgaria, Ukr. Food J. 5 (2016) 492–498.

[106] S. Price, Aromatherapy for Common Ailments: How to Use Essential Oils—Such as Rosemary, Chamomile, and Lavender—to Prevent and Treat More than 40 Common Ailments, Simon and Schuster, Inc, New York, USA, 2003, pp. 1–93.

[107] E. Fradelos, A. Komini, The use of essential oils as a complementary treatment for anxiety, Am. J. Nurs. Sci. 4 (2015) 1–5.

[108] A.C. de Groot, E. Schmidt, Tea tree oil: contact allergy and chemical composition, Contact Dermatitis 75 (2016) 129–143.

[109] C.F. Carson, T.V. Riley, Susceptibility of Propionibacterium acnes to the essential oil of Melaleuca alternifolia, Lett. Appl. Microbiol. 19 (1994) 24–25.

[110] C.F. Carson, K.A. Hammer, T.V. Riley, Melaleuca alternifolia (tea tree) oil: a review of antimicrobial and other medicinal properties, Clin. Microbiol. Rev. 19 (2006) 50–62.

[111] K. Hammer, C.F. Carson, T.V. Riley, Antifungal activity of the components of Melaleuca alternifolia (tea tree) oil, J. Appl. Microbiol. 95 (2003) 853–860.

[112] U. Muchjajib, S. Muchjajib, Effect of picking time on essential oil yield of Ylang-Ylang (Cananga odorata), in: IXXVIII International Horticultural Congress on Science and Horticulture for People (IHC2010): A New Look at Medicinal and Aromatic Plants Seminar, Lisbon, Portugal, 2010, pp. 243–248.

[113] N. Bleasel, B. Tate, M. Rademaker, Allergic contact dermatitis following exposure to essential oils, Aust. J. Dermatol. 43 (2002) 211–213.

[114] A. Wei, T. Shibamoto, Antioxidant activities and volatile constituents of various essential oils, J. Agric. Food Chem. 55 (2007) 1737–1742.

[115] C.K. Huang, T.A. Miller, The truth about over-the-counter topical anti-aging products: a comprehensive review, Aesthet. Surg. J. 27 (2007) 402–412.

[116] J.R. Thomas, T.K. Dixon, T.K. Bhattacharyya, Effects of topicals on the aging skin process, Facial Plast. Surg. Cl. 21 (2013) 55–60.

[117] S. Sonti, E.T. Makino, J.A. Garruto, J.V. Gruber, S. Rao, R.C. Mehta, Efficacy of a novel treatment serum in the improvement of photodamaged skin, Int. J. Cosmet. Sci. 35 (2013) 156–162.

[118] S.K. Schagen, Topical peptide treatments with effective anti-aging results, Cosmetics 4 (2017) 16–30.

[119] H. Sundaram, A. Cegielska, A. Wojciechowska, P. Delobel, Prospective, randomized, investigator-blinded, split-face evaluation of a topical crosslinked hyaluronic acid serum for post-procedural improvement of skin quality and biomechanical attributes, J. Drugs Dermatol. 17 (2018) 442–450.

[120] H. Nazar, A. Nazar, M. Nazar, Beauty is now more than skin deep-the emergence of cosmeceuticals, Pharm. J. 292 (2014) 380.

[121] S. Sunder, Relevant topical skin care products for prevention and treatment of aging skin, Facial Plast. Surg. Cl. 27 (2019) 413–418.

[122] Y. Naraoka, A. Hu, T. Yamaguchi, N. Saga, H. Kobayashi, 5-Aminolevulinic acid improves water content and reduces skin wrinkling, Health 12 (2020) 709–716.

[123] P.C. Chiu, C.C. Chan, H.M. Lin, H.C. Chiu, The clinical anti-aging effects of topical kinetin and niacinamide in Asians: a randomized, double-blind, placebo-controlled, split-face comparative trial, J. Cosmet. Dermatol. 6 (2007) 243–249.

[124] E. Raposio, V. Belgrano, A. Caielli, E. Canini, E. Grosso, V. Pavacci, C. Porzio, C. Rossello, P. Santi, Evaluation of effectiveness of viper serum for topical use as facial anti-aging, Capsula Eburnean 4 (2009) 1–6.

[125] R. Gopaul, H.E. Knaggs, J.F. Lephart, K.C. Holley, E.M. Gibson, Original contribution: an evaluation of the effect of a topical product containing salicin on the visible signs of human skin aging, J. Cosmet. Dermatol. 9 (2010) 196–201.

[126] M.T. Sanz, C. Campos, M. Milani, M. Foyaca, A. Lamy, K. Kurdian, C. Trullas, Biorevitalizing effect of a novel facial serum containing apple stem cell extract, pro-collagen lipopeptide, creatine, and urea on skin aging signs, J. Cosmet. Dermatol. 15 (2016) 24–30.

[127] C. Aldag, D.N. Teixeira, P.S. Leventhal, Skin rejuvenation using cosmetic products containing growth factors, cytokines, and matrikines: a review of the literature, Clin. Cosmet. Investig. Dermatol. 9 (2016) 411–419.

[128] S.G. Fabi, L. Zaleski-Larsen, J. Bolton, R.C. Mehta, E.T. Makino, Optimizing facial rejuvenation with a combination of a novel topical serum and injectable procedure to increase patient outcomes and satisfaction, J. Clin. Aesthet. Dermatol. 10 (2017) 14–18.

[129] N.N. Suhendra, The efficacy of 7% palmitoyl pentapeptide-4 serum for the periorbital wrinkle reduction, in: Rangsit Graduate Research Conference, Bangkok, Thailand, 2020, pp. 2870–2884.

[130] V. Chowjarean, P.P. Phiboonchaiyanan, S. Harikarnpakdee, P. Tengamnuay, A natural skin anti-ageing serum containing pseudobulb ethanolic extract of Grammatophyllum speciosum: a randomized double-blind, placebo-controlled trial, Int. J. Cosmet. Sci. 41 (2019) 548–557.

[131] N. Lourith, M. Kanlayavattanakul, Formulation and clinical evaluation of the standardized Litchi chinensis extract for skin hyperpigmentation and aging treatments, Ann. Pharm. Fr. 78 (2020) 142–149.

[132] D. Bacqueville, A. Maret, M. Noizet, L. Duprat, C. Coutanceau, V. Georgescu, S. Bessou-Touya, H. Duplan, Efficacy of a dermocosmetic serum combining bakuchiol and vanilla tahitensis extract to prevent skin photoaging in vitro and to improve clinical outcomes for naturally aged skin, Clin. Cosmet. Investig. Dermatol. 13 (2020) 359–370.

[133] L. Mungmai, W. Preedalikit, N. Aunsri, D. Amornlerdpison, Efficacy of cosmetic formulation containing Perilla frutescens leaves extract for irritation and aging skin, Biomed. Pharmacol. J. 13 (2020) 779–787.

[134] P. Rattanawiwatpong, R. Wanitphakdeedecha, A. Bumrungpert, M. Maiprasert, Anti-aging and brightening effects of a topical treatment containing vitamin C, vitamin E, and raspberry leaf cell culture extract: a split-face, randomized controlled trial, J. Cosmet. Dermatol. 19 (2020) 671–676.

[135] Y. Harth, M. Rapoport, Photosensitivity associated with antipsychotics, antidepressants and anxiolytics, Drug Saf. 14 (1996) 252–259.

[136] S.A. Jabbour, Cutaneous manifestations of endocrine disorders, Am. J. Clin. Dermatol. 4 (2003) 315–331.

[137] R.M. Halder, G.M. Richards, Topical agents used in the management of hyperpigmentation, Skin Therapy Lett. 9 (2004) 1–3.

[138] A.N. Geria, A.L. Tajirian, G. Kihiczak, R.A. Schwartz, Minocycline-induced skin pigmentation: an update, Acta Dermatovenerol. Croat. 17 (2009) 123–126.

[139] E. Makino, T. Priscilla, M. Rahul, Cosmetic efficacy of a comprehensive serum for facial hyperpigmentation in multiple ethnic populations, J. Am. Acad. Dermatol. 76 (2017) 85–89.

[140] D.A. Dickinson, H.J. Forman, Glutathione in defense and signaling: lessons from a small thiol, Ann. N. Y. Acad. Sci. 973 (2002) 488–504.

[141] R. Sarkar, S. Chugh, V.K. Garg, Newer and upcoming therapies for melasma, Indian J. Dermatol. Venereol. Leprol. 78 (2012) 417–428.

[142] C. Couteau, L. Coiffard, Overview of skin whitening agents: drugs and cosmetic products, Cosmetologica 3 (2016) 27–30.

[143] S. Sonthalia, D. Daulatabad, R. Sarkar, Glutathione as a skin whitening agent: facts, myths, evidence and controversies, Indian J. Dermatol. Venereol. Leprol. 82 (2016) 262–272.

[144] Y.C. Boo, Human skin lightening efficacy of resveratrol and its analogs: from in vitro studies to cosmetic applications, Antioxidants 8 (2019) 332–349.

[145] K.M. Babbush, R.A. Babbush, A. Khachemoune, Treatment of melasma: a review of less commonly used antioxidants, Int. J. Dermatol. 59 (2020) 1–8.

[146] S. Mohan, L. Mohan, R. Sangal, N. Singh, Glutathione for skin lightening for dermatologists and cosmetologists, Int. J. Res. 6 (2020) 284–287.

[147] M. Narda, A. Brown, B. Muscatelli-Groux, J.A. Grimaud, C. Granger, Epidermal and dermal hallmarks of photoaging are prevented by treatment with night serum containing melatonin, bakuchiol, and ascorbyl tetraisopalmitate: in vitro and ex vivo studies, Dermatol. Ther. 10 (2020) 191–202.

[148] V.M. Verallo-Rowell, V. Verallo, K. Graupe, L. Lopez-Villafuerte, M. Garcia-Lopez, Double-blind comparison of azelaic acid and hydroquinone in the treatment of melasma, Acta Derm. Venereol. Suppl. 143 (1989) 58–61.

[149] K. Jimbow, N-acetyl-4-S-cysteaminylphenol as a new type of depigmenting agent for the melanoderma of patients with melasma, Arch. Dermatol. 127 (1991) 1528–1534.

[150] M. Nakagawa, K. Kawai, K. Kawai, Contact allergy to kojic acid in skin care products, Contact Dermatitis 32 (1995) 9–13.

[151] F.H. Lin, J.Y. Lin, R.D. Gupta, J.A. Tournas, J.A. Burch, M.A. Selim, N.A. Monteiro-Riviere, J.M. Grichnik, J. Zielinski, S.R. Pinnell, Ferulic acid stabilizes a solution of vitamins C and E and doubles its photoprotection of skin, J. Invest. Dermatol. 125 (2005) 826–832.

[152] J.E. Zielinski, S.R. Pinnell, Stabilized ascorbic acid compositions and methods thereof, The United States Patent 7,179,841, 2007.

[153] A. Khemis, A. Kaiafa, C. Queille-Roussel, L. Duteil, J.P. Ortonne, Evaluation of efficacy and safety of rucinol serum in patients with melasma: a randomized controlled trial, Br. J. Dermatol. 156 (2007) 997–1004.

[154] T. Amnuaikit, S. Khakhong, P. Khongkow, Formulation development and facial skin evaluation of serum containing jellose from tamarind seeds, J. Pharm. Res. Int. 31 (2019) 1–14.

[155] S. Desai, E. Ayres, H. Bak, et al., Effect of a tranexamic acid, kojic acid, and niacinamide containing serum on facial dyschromia: a clinical evaluation, J. Drugs Dermatol. 18 (2019) 454–459.

[156] B.L. Kedrowski, M.J. Mueller, Skin lightening compounds from fruit seed extracts, The United States Patent 16/322,241, 2019.

[157] E. Bronzina, A. Clement, B. Marie, K.T. Fook Chong, P. Faure, T. Passeron, Efficacy and tolerability on melasma of a topical cosmetic product acting on melanocytes, fibroblasts and endothelial cells: a randomized comparative trial against 4% hydroquinone, J. Eur. Acad. Dermatol. Venereol. 34 (2020) 897–903.

[158] K.A. Arndt, T.B. Fitzpatrick, Topical use of hydroquinone as a depigmenting agent, JAMA 194 (1965) 965–967.

[159] T. Piamphongsant, Treatment of melasma: a review with personal experience, Int. J. Dermatol. 37 (1998) 897–903.

[160] S.B. Ennes, R.C. Paschoalick, M.M. Alchorne, A double-blind, comparative, placebo-controlled study of the efficacy and tolerability of 4% hydroquinone as a depigmenting agent in melasma, J. Dermatol. Treat. 11 (2000) 173–179.

[161] M. Rendon, M. Berneburg, I. Arellano, M. Picardo, Treatment of melasma, J. Am. Acad. Dermatol. 54 (2006) S272–S281.

[162] R.C. Kelm, A.S. Zahr, T. Kononov, O. Ibrahim, Effective lightening of facial melasma during the summer with a dual regimen: a prospective, open-label, evaluator-blinded study, J. Cosmet. Dermatol. 19 (2020) 1–21.

[163] L. Atzori, M.A. Brundu, A. Orru, P. Biggio, Glycolic acid peeling in the treatment of acne, J. Eur. Acad. Dermatol. Venereol. 12 (1999) 119–122.

[164] H.S. Lee, I.H. Kim, Salicylic acid peels for the treatment of acne vulgaris in Asian patients, Dermatol. Surg. 29 (2003) 1196–1199.

[165] S. Gupta, S.K. Gupta, Hydroxy acids based delivery systems for skin resurfacing and anti-aging compositions, United States Patent 10/290,933, 2004.

[166] W. Walden, Oil-based composition for acne, The United States Patent 11/293,692, 2006.

[167] E. Kessler, K. Flanagan, C. Chia, C. Rogers, G.D. Anna, Comparison of α-and β-hydroxy acid chemical peels in the treatment of mild to moderately severe facial acne vulgaris, Dermatol. Surg. 34 (2008) 45–51.

[168] A.L. Zaenglein, A.L. Pathy, B.J. Schlosser, A. Alikhan, H.E. Baldwin, D.S. Berson, W.P. Bowe, E.M. Graber, J.C. Harper, S. Kang, J.E. Keri, Guidelines of care for the management of acne vulgaris, J. Am. Acad. Dermatol. 74 (2016) 945–973.

[169] T.X. Cong, D. Hao, X. Wen, et al., From pathogenesis of acne vulgaris to anti-acne agents, Arch. Dermatol. Res. 311 (2019) 337–349.

[170] K. Lee, M.S. Shin, I. Ham, H.Y. Choi, Investigation of the mechanisms of Angelica dahurica root extract-induced vasorelaxation in isolated rat aortic rings, BMC Complement. Altern. Med. 15 (2015) 1–8.

[171] C. Nam, S. Kim, Y. Sim, I. Chang, Anti-acne effects of oriental herb extracts: a novel screening method to select anti-acne agents, Skin Pharmacol. Physiol. 16 (2003) 84–90.

[172] S.M. Han, K.G. Lee, S.C. Pak, Effects of cosmetics containing purified honeybee (Apis mellifera L.) venom on acne vulgaris, J. Integr. Med. 11 (2013) 320–326.

[173] S.M. Han, S.C. Pak, Y.M. Nicholls, N. Macfarlane, Evaluation of anti-acne property of purified bee venom serum in humans, J. Cosmet. Dermatol. 15 (2016) 324–329.

[174] N. Fitri, I. Fatimah, L. Chabib, F.I. Fajarwati, Formulation of antiacne serum based on lime peel essential oil and in vitro antibacterial activity test against Propionibacterium acnes, in: AIP Conference Proceedings, Yogyakarta, Indonesia, 2017, pp. 0201231–0201237.

[175] M.P. Jagtap, M.V. Chaudhari, M.R. Davar, M.N. Patil, M.P. Joshi, M.B. Desale, Formulation and development of anti-acne serum using Euphorbia hirta, Int. J. Creat. Innov. Res. Stud. 2 (2020) 171–179.

[176] T. Gajjar, N. Patel, Safety and efficacy of two botanical based topical anti-acne products in treatment of mild to moderate acne subjects, J. Dermat. Cosmetol. 4 (2020) 99–107.

[177] G. Chernoff, The utilization of a topical nitric oxide generating serum in aesthetic medicine, J. Dermatol. Surg. 5 (2020) 1329–1340.

[178] G. Chernoff, The utilization of a nitric oxide generating serum in the treatment of active acne and acne scarred patients, J. Biomed. Res. Environ. Sci. 6 (2020) 010–014.

[179] S.N. Bukhari, N.L. Roswandi, M. Waqas, H. Habib, F. Hussain, S. Khan, M. Sohail, N.A. Ramli, H.E. Thu, Z. Hussain, Hyaluronic acid, a promising skin rejuvenating biomedicine: a review of recent updates and pre-clinical and clinical investigations on cosmetic and nutricosmetic effects, Int. J. Biol. Macromol. 120 (2018) 1682–1695.

[180] S. Schwartz, E. Frank, D. Gierhart, P. Simpson, R. Frumento, Zeaxanthin-based dietary supplement and topical serum improve hydration and reduce wrinkle count in female subjects, J. Cosmet. Dermatol. 15 (2016) e13–e20.

[181] E. Anitua, M. Troya, A. Pino, A novel protein-based autologous topical serum for skin regeneration, J. Cosmet. Dermatol. 19 (2020) 705–713.

[182] R.A. Lodge, B. Bhushan, Effect of physical wear and triboelectric interaction on surface charge as measured by Kelvin probe microscopy, J. Colloid Interface Sci. 310 (2007) 321–330.

[183] S.C. Mehta, P. Somasundaran, Mechanism of stabilization of silicone oil–water emulsions using hybrid siloxane polymers, Langmuir 24 (2008) 4558–4563.

[184] L.A. Araújo, F. Addor, P.M. Campos, Use of silicon for skin and hair care: an approach of chemical forms available and efficacy, An. Bras. Dematol. 91 (2016) 331–335.

[185] J.V. Gruber, B.R. Lamoureux, N. Joshi, L. Moral, Influence of cationic polysaccharides on polydimethylsiloxane (PDMS) deposition onto keratin surfaces from a surfactant emulsified system, Colloids Surf. B Biointerfaces 19 (2000) 127–135.

[186] C.T. La, B. Bhushan, Nanotribological effects of silicone type, silicone deposition level, and surfactant type on human hair using atomic force microscopy, J. Cosmet. Sci. 57 (2006) 37–56.

[187] H. Nazir, L. Wang, G. Lian, S. Zhu, Y. Zhang, Y. Liu, G. Ma, Multilayered silicone oil droplets of narrow size distribution: preparation and improved deposition on hair, Colloids Surf. B Biointerfaces 100 (2012) 42–49.

[188] H. Nazir, W. Zhang, Y. Liu, X. Chen, L. Wang, M.M. Naseer, G. Ma, Silicone oil emulsions: strategies to improve their stability and applications in hair care products, Int. J. Cosmet. Sci. 36 (2014) 124–133.

[189] A.H. Rashaid, P.B. Harrington, G.P. Jackson, Amino acid composition of human scalp hair as a biometric classifier and investigative lead, Anal. Methods 7 (2015) 1707–1718.

[190] P. Tessari, A. Lante, G. Mosca, Essential amino acids: master regulators of nutrition and environmental footprint? Sci. Rep. 6 (2016) 26074.

[191] D. Bouilly-Gauthier, C. Jeannes, N. Dupont, N. Piccardi, P. Manissier, U. Heinrich, H. Tronnier, A new nutritional supplementation is effective against hair loss and improves hair quality, in: European Congress on AntiAging and Aesthetic Medicine, Paris, France, 2008.

[192] E. Haneke, R. Baran, Micronutrients for hair and nails, in: J. Krutmann, P. Humbert (Eds.), Nutrition for Healthy Skin, Springer, Berlin, Germany, 2010, pp. 149–163.

[193] R. Saini, S.L. Badole, A.A. Zanwar, Arginine derived nitric oxide: key to healthy skin, in: R. Watson, S. Zibadi (Eds.), Bioactive Dietary Factors and Plant Extracts in Dermatology, Humana Press, Totowa, NJ, 2013, pp. 73–82.

[194] D.H. Rushton, Nutritional factors and hair loss, Clin. Exp. Dermatol. 27 (2002) 396–404.

[195] S. Méndez, A.M. Manich, M. Martí, J.L. Parra, L. Coderch, Damaged hair retrieval with ceramide-rich liposomes, J. Cosmet. Sci. 62 (2011) 565–577.

[196] B.M. Park, S.S. Bak, K.O. Shin, M. Kim, D. Kim, S.H. Jung, S. Jeong, Y.K. Sung, H.J. Kim, Promotion of hair growth by newly synthesized ceramide mimetic compound, Biochem. Biophys. Res. Commun. 491 (2017) 173–177.

[197] S. Eom, H.J. Hyun, H.Y. Kim, A study on the hair care formulations containing various amounts of ceramide as anti-aging cosmetics, in: Joint Event on 8th International Conference on Cosmetology & Skin Care & 14th International Conference and Exhibition on Cosmetic Dermatology and Hair Care, Madrid, Spain, 2018.

[198] E. Kahraman, M. Kaykın, H. Şahin Bektay, S. Güngör, Recent advances on topical application of ceramides to restore barrier function of skin, Cosmetics 6 (2019) 52–62.

[199] A.W. Rafi, R.M. Katz, Pilot study of 15 patients receiving a new treatment regimen for androgenic alopecia: the effects of atopy on AGA, ISRN Dermatol. 2011 (2011) 241953.

[200] S. Jose, M.L. Hughbanks, B.Y. Binder, G.C. Ingavle, J.K. Leach, Enhanced trophic factor secretion by mesenchymal stem/stromal cells with glycine-histidine-lysine (GHK)-modified alginate hydrogels, Acta Biomater. 10 (2014) 1955–1964.

[201] A. Monselise, D.E. Cohen, R. Wanser, J. Shapiro, What ages hair? Int. J. Women's Dermatol. 3 (2017) S52–S57.

[202] T.W. Fischer, U.C. Hipler, P. Elsner, Effect of caffeine and testosterone on the proliferation of human hair follicles in vitro, Int. J. Dermatol. 46 (2007) 27–35.

[203] L. Montenegro, G. Puglisi, Evaluation of sunscreen safety by in vitro skin permeation studies: effects of vehicle composition, Pharmazie 68 (2013) 34–40.

[204] M. Majeed, S. Majeed, K. Nagabhushanam, L. Mundkur, P. Neupane, K. Shah, Clinical study to evaluate the efficacy and safety of a hair serum product in healthy adult male and female volunteers with hair fall, Clin. Cosmet. Investig. Dermatol. 13 (2020) 691–700.

[205] X. Hui, S.B. Hornby, R.C. Wester, S. Barbadillo, Y. Appa, H. Maibach, In vitro human nail penetration and kinetics of panthenol, Int. J. Cosmet. Sci. 29 (2007) 277–282.

[206] U. Runne, C.E. Orfanos, The human nail, in: J.W.H. Mali (Ed.), Some Fundamental Approaches in Skin Research, Karger Publisher, Basel, Switzerland, 1981, pp. 102–149.

[207] A.R. Rathi, R.R. Popat, V.S. Adhao, V.N. Shrikhande, Nail drug delivery system: a review, Int. J. Pharm. Chem. Anal. 7 (2020) 9–21.

[208] H.B. Gunt, G.B. Kasting, Effect of hydration on the permeation of ketoconazole through human nail plate in vitro, Eur. J. Pharm. Sci. 32 (2007) 254–260.

[209] P. Shende, D. Patel, A. Takke, Nanomaterial-based cosmeceuticals, in: C.M. Hussain (Ed.), Handbook of Functionalized Nanomaterials for Industrial Applications, Elsevier, USA, 2020, pp. 775–791.

[210] H.C. Williams, R. Buffham, A. du Vivier, Successful use of topical vitamin E solution in the treatment of nail changes in yellow nail syndrome, Arch. Dermatol. 127 (1991) 1023–1028.

[211] P.J. DiMeglio, Nail oil composition, United States Patent 4,810,498, 1989.

[212] T. Rockhill, Antifungal serum, The United States Patent 9,439,972, 2016.

[213] L. Braguti, J.A. Randall, Novel antifungal compositions, The United States Patent 16/629,975, 2020.

[214] F. Fernandez-Campos, F. Navarro, A. Corrales, J. Picas, E. Pena, J. González, F.J. Otero-Espinar, Transungual delivery, anti-inflammatory activity, and in vivo assessment of a cyclodextrin polypseudorotaxanes nail lacquer, Pharmaceutics 12 (2020) 730.

[215] N.D. Scancarella, J.A. Duffy, M.S. Garrison, G.K. Menon, Composition and method for under-eye skin lightening, The United States Patent 5,643,587, 1997.

[216] T. Mitsuishi, T. Shimoda, Y. Mitsui, Y. Kuriyama, S. Kawana, The effects of topical application of phytonadione, retinol and vitamins C and E on infraorbital dark circles and wrinkles of the lower eyelids, J. Cosmet. Dermatol. 3 (2004) 73–75.

[217] F. Gorouhi, H.I. Maibach, Role of topical peptides in preventing or treating aged skin, Int. J. Cosmet. Sci. 31 (2009) 327–345.

[218] E. Papakonstantinou, M. Roth, G. Karakiulakis, Hyaluronic acid: a key molecule in skin aging, Dermato-Endocrinol. 4 (2012) 253–258.

[219] S. Silva, M. Ferreira, A.S. Oliveira, C. Magalhaes, M.E. Sousa, M. Pinto, J.M. Sousa Lobo, I.F. Almeida, Evolution of the use of antioxidants in anti-ageing cosmetics, Int. J. Cosmet. Sci. 41 (2019) 378–386.

[220] D. Cohen, M. Portugal-Cohen, Safe retinol-like skin biological effect by a new complex, enriched with retinol precursors, J. Cosmet. Dermatol. Sci. Appl. 10 (2020) 59–65.

[221] S. Jeong, S. Yoon, S. Kim, J. Jung, M. Kor, K. Shin, C. Lim, H.S. Han, H. Lee, K.Y. Park, J. Kim, Anti-wrinkle benefits of peptides complex stimulating skin basement membrane proteins expression, Int. J. Mol. Sci. 21 (2020) 73.

[222] M. Kadu, S. Vishwasrao, S. Singh, Review on natural lip balm, Int. J. Res. Cosmet. Sci. 5 (2015) 1–7.

[223] N.S. Trookman, R.L. Rizer, R. Ford, R. Mehta, V. Gotz, Clinical assessment of a combination lip treatment to restore moisturization and fullness, J. Clin. Aesther. Dermatol. 2 (2009) 44–48.

[224] K. Lintner, Cosmetic or dermopharmaceutical use of peptides for healing, hydrating and improving skin appearance during natural or induced ageing (heliodermia, pollution), United States Patent 6,620,419, 2003.

[225] H.I. Maibach, E.C. Luo, T.M. Hsu, Method and topical formulation for treating skin conditions associated with aging, The United States Patent 7,205,003, 2007.

[226] L. Zhang, T.J. Falla, Cosmeceuticals and peptides, Clin. Dermatol. 27 (2009) 485–494.

[227] S.M. Raymond-Coblantz, Hand and body moisturizing serum, The United States Patent 13/506,982, 2013.

[228] J.N. Campbell, How does topical lidocaine relieve pain? Pain 153 (2012) 255–256.

[229] G.L. Silva, C. Luft, A. Lunardelli, R.H. Amaral, D.A. Melo, M.V. Donadio, F.B. Nunes, M.S. de Azambuja, J.C. Santana, C. Moraes, R.O. Mello, E. Cassel, M.A. Periera, J.R. de Oliveira, Antioxidant, analgesic and anti-inflammatory effects of lavender essential oil, An. Acad. Bras. Cienc. 87 (2015) 1397–1408.

[230] B. Ali, N.A. Al-Wabel, S. Shams, A. Ahamad, S.A. Khan, F. Anwar, Essential oils used in aromatherapy: a systemic review, Asian Pac. J. Trop. Biomed. 5 (2015) 601–611.

[231] M.S. Gerstel, V.A. Place, Drug delivery device, United States patent 3,964,482, 1976.

[232] S. Doddaballapur, Microneedling with dermaroller, J. Cutan. Aesthet. Surg. 2 (2009) 110–111.

[233] D. Bhardwaj, Collagen induction therapy with dermaroller, Commun Based Med J. 1 (2013) 35–37.

[234] M.T. McCrudden, E. McAlister, A.J. Courtenay, P. González-Vázquez, T.R. Raj Singh, R.F. Donnelly, Microneedle applications in improving skin appearance, Exp. Dermatol. 24 (2015) 561–566.

[235] D. Fernandes, Current concepts on how to optimise skin needling 2020: a personal experience: part 2, Dermatol. Rev. 1 (2020) 10–15.

[236] G. Fabbrocini, V. De Vita, L. Di Costanzo, F. Pastore, M.C. Mauriello, M. Ambra, M.C. Annunziata, M.G. di Santolo, N. Cameli, G. Monfrecola, Skin needling in the treatment of the aging neck, Skinmed 9 (2011) 347–351.

[237] G. Fabbrocini, V. De Vita, N. Fardella, F. Pastore, M.C. Annunziata, M.C. Mauriello, A. Monfrecola, N. Cameli, Skin needling to enhance depigmenting serum penetration in the treatment of melasma, Plast. Surg. Int. 2011 (2011) 1–7.

[238] A. Singh, S. Yadav, Microneedling: advances and widening horizons, Indian Dematol. Online J. 7 (2016) 244–254.

[239] S. Bhatnagar, K. Dave, V.V. Venuganti, Microneedles in the clinic, J. Control. Release 260 (2017) 164–182.

[240] S.L. Beergouder, A. Reshme, Scalp micro-needling: a new tool in the treatment of alopecia totalis, Clin. Dermatol. Rev. 4 (2020) 164–166.

[241] G. Fabbrocini, C. Mazzella, M. D'Andrea, Microneedling for neocollagenesis of the face, in: A. Costa (Ed.), Minimally Invasive Aesthetic Procedures, Springer, Cham, Switzerland, 2020, pp. 625–630.

[242] A. Villani, M. Carmela Annunziata, M. Antonietta Luciano, G. Fabbrocini, Skin needling for the treatment of acne scarring: a comprehensive review, J. Cosmet. Dermatol. 19 (2020) 2174–2181.

[243] G. Yang, G. Chen, Z. Gu, Transdermal drug delivery for hair regrowth, Mol. Pharm. 17 (2020) 1–8.

[244] P.A. Nair, T.H. Arora, Microneedling using dermaroller: a means of collagen induction therapy, Galen. Med. J. 69 (2014) 24–27.

[245] A.J. Yu, Y.J. Luo, X.G. Xu, L.L. Bao, T. Tian, Z.X. Li, Y.X. Dong, Y.H. Li, A pilot split-scalp study of combined fractional radiofrequency microneedling and 5% topical minoxidil in treating male pattern hair loss, Clin. Exp. Dermatol. 43 (2018) 775–781.

[246] K.B. Shah, A.N. Shah, R.B. Solanki, R.C. Raval, A comparative study of microneedling with platelet-rich plasma plus topical minoxidil (5%) and topical minoxidil (5%) alone in androgenetic alopecia, Int. J. Trichol. 9 (2017) 14–18.

[247] B.S. Chandrashekar, V. Yepuri, V. Mysore, Alopecia areata-successful outcome with microneedling and triamcinolone acetonide, J. Cutan. Aesthet. Surg. 7 (2014) 63–64.

[248] W. Alsalhi, A. Alalola, M. Randolph, E. Gwillim, A. Tosti, Novel drug delivery approaches for the management of hair loss, Expert Opin. Drug Deliv. 17 (2020) 287–295.

[249] M. El Mofty, S. Esmat, N. Hunter, H.M. Mashaly, D. Dorgham, O. Shaker, S. Ibrahim, Effect of different types of therapeutic trauma on vitiligo lesions, Dermatol. Ther. 30 (2017) e12447.

[250] E.M. Attwa, S.A. Khashaba, N.A. Ezzat, Evaluation of the additional effect of topical 5-fluorouracil to needling in the treatment of localized vitiligo, J. Cosmet. Dermatol. 19 (2020) 1473–1478.

[251] E.V. Lima, M.M. Lima, H.A. Miot, Induction of pigmentation through microneedling in stable localized vitiligo patients, Dermatol. Surg. 46 (2020) 434–435.

[252] A. Salloum, N. Bazzi, D. Maalouf, M. Habre, Microneedling in vitiligo: a systematic review, Dermatol. Ther. 33 (2020) e14297.

[253] A.K. Jha, S. Sonthalia, 5-Fluorouracil as an adjuvant therapy along with microneedling in vitiligo, J. Am. Acad. Dermatol. 80 (2019) e75–e76.

[254] A. Kumar, R. Bharti, S. Agarwal, Microneedling with dermaroller 192 needles along with 5-fluorouracil solution in the treatment of stable vitiligo, J. Am. Acad. Dermatol. 81 (2019) e67–e69.

[255] M. Mina, L. Elgarhy, H. Al-Saeid, Z. Ibrahim, Comparison between the efficacy of microneedling combined with 5-fluorouracil vs microneedling with tacrolimus in the treatment of vitiligo, J. Cosmet. Dermatol. 17 (2018) 744–751.

[256] V. Mirela, M.S. Elena, P. Dejan, G. Lazarova, P.N. Biljana, B.B. Vesna, D. Irena, Efficacy of 5 fluorouracil solution application using dermaroller in the treatment of vitiligo, Int. J. Med. Clin. Res. Rev. 3 (2020) 290–297.

[257] F.T. Zahra, M. Adil, S.S. Amin, M. Mohtashim, R. Bansal, H.Q. Khan, Efficacy of topical 5% 5-fluorouracil with needling versus 5% 5-fluorouracil alone in stable vitiligo: a randomized controlled study, J. Cutan. Aesthet. Surg. 13 (2020) 197–203.

[258] I.V. Korobko, K.M. Lomonosov, A pilot comparative study of topical latanoprost and tacrolimus in combination with narrow-band ultraviolet B phototherapy and microneedling for the treatment of nonsegmental vitiligo, Dermatol. Ther. 29 (2016) 437–441.

[259] S.A. Elshafy Khashaba, R.A. Elkot, A.M. Ibrahim, Efficacy of NB-UVB, microneedling with triamcinolone acetonide, and a combination of both modalities in the treatment of vitiligo: a comparative study, J. Am. Acad. Dermatol. 79 (2018) 365–367.

[260] Z.A. Ibrahim, G.F. Hassan, H.Y. Elgendy, H.A. Al-shenawy, Evaluation of the efficacy of transdermal drug delivery of calcipotriol plus betamethasone versus tacrolimus in the treatment of vitiligo, J. Cosmet. Dermatol. 18 (2019) 581–588.

[261] H.M. Ebrahim, W. Albalate, Efficacy of microneedling combined with tacrolimus versus either one alone for vitiligo treatment, J. Cosmet. Dermatol. 19 (2020) 855–862.

[262] M. Khater, M. Nasr, S. Salah, F. Khattab, Clinical evaluation of the efficacy of trichloroacetic acid 70% after micro-needling vs intradermal injection of 5-fluorouracil in the treatment of non-segmental vitiligo; a prospective comparative study, Dermatol. Ther. 33 (2020) e13532.

[263] M. El-Domyati, N.H. Moftah, G.A. Nasif, S.W. Ameen, M.R. Ibrahim, M.H. Ragaie, Facial rejuvenation using stem cell conditioned media combined with skin needling: a split-face comparative study, J. Cosmet. Dermatol. 19 (2020) 2404–2410.

[264] K.Y. Park, H.J. Kwon, C. Lee, D. Kim, J.J. Yoon, M.N. Kim, B.J. Kim, Efficacy and safety of a new microneedle patch for skin brightening: a randomized, split-face, single-blind study, J. Cosmet. Dermatol. 16 (2017) 382–387.

[265] M. Avcil, G. Akman, J. Klokkers, D. Jeong, A. Çelik, Efficacy of bioactive peptides loaded on hyaluronic acid microneedle patches: a monocentric clinical study, J. Cosmet. Dermatol. 19 (2020) 328–337.

[266] J.F. Ramírez-Oliveros, L. de Abreu, C. Tamler, P. Vilhena, M.H. de Barros, Microneedling with drug delivery (hydroquinone 4% serum) as an adjuvant therapy for recalcitrant melasma, Skinmed 18 (2020) 38–40.

[267] L. Budamakuntla, E. Loganathan, D.H. Suresh, S. Shanmugam, S. Suryanarayan, A. Dongare, L.D. Venkataramiah, N. Prabhu, A randomised, open-label, comparative study of tranexamic acid microinjections and tranexamic acid with microneedling in patients with melasma, J. Cutan. Aesthet. Surg. 6 (2013) 139–143.

[268] E.R. Hofny, A.A. Abdel-Motaleb, A. Ghazally, A.M. Ahmed, M.R. Hussein, Platelet-rich plasma is a useful therapeutic option in melasma, J. Dermatolog. Treat. 30 (2019) 396–401.

[269] M.M. Gamea, D.A. Kamal, A.A. Donia, D.S. Hegab, Comparative study between topical tranexamic acid alone versus its combination with autologous platelet rich plasma for treatment of melasma, J. Dermatolog. Treat. 31 (2020) 1–7.

[270] A. Menon, H. Eram, P.R. Kamath, S. Goel, A.M. Babu, A split face comparative study of safety and efficacy of microneedling with tranexamic acid versus microneedling with vitamin C in the treatment of melasma, Indian Dermatol. Online J. 11 (2020) 41–45.

[271] A.A. Aly, R.E. AbdElMaksoud, I. Ismail, Tranexamic acid versus topical mesolightening mixture using the dermaroller in the treatment of melisma, Egypt. J. Dermatol. Venerol. 40 (2020) 45–52.

[272] M.J. Pikal, The role of electroosmotic flow in transdermal iontophoresis, Adv. Drug Deliv. Rev. 46 (2001) 281–305.

[273] Y.N. Kalia, A. Naik, J. Garrison, R.H. Guy, Iontophoretic drug delivery, Adv. Drug Deliv. Rev. 56 (2004) 619–658.

[274] S. Rawat, S. Vengurlekar, B. Rakesh, S. Jain, G. Srikarti, Transdermal delivery by iontophoresis, Indian J. Pharm. Sci. 70 (2008) 5–10.

[275] X. Ning, R.Z. Seeni, C. Xu, Iontophoresis enhanced transdermal drug delivery, in: C. Xu, X. Wang, M. Prammanik (Eds.), Imaging Technologies and Transdermal Delivery in Skin Disorders, Wiley-VCH Verlag GmbH & Co., Germany, 2019, pp. 291–307.

[276] P. Bakshi, D. Vora, K. Hemmady, A.K. Banga, Iontophoretic skin delivery systems: success and failures, Int. J. Pharm. 586 (2020) 119584.

[277] N. Dixit, V. Bali, S. Baboota, A. Ahuja, J. Ali, Iontophoresis—an approach for controlled drug delivery: a review, Curr. Drug Deliv. 4 (2007) 1–10.

[278] Y. Sun, J. Wu, J.C. Liu, J. Chantalat, Methods of reducing the appearance of pigmentation with galvanic generated electricity, United States Patent 7,476,222, 2009.

[279] A. Khan, M. Yasir, M. Asif, I. Chauhan, A. Singh, R. Sharma, et al., Iontophoretic drug delivery: history and applications, J. Appl. Pharm. Sci. 1 (2011) 11–24.

[280] M.S.L. Datuin, Iontophoresis and mesotherapy in melasma, in: E. Handog, M. Enriquez-Macarayo (Eds.), Melasma and Vitiligo in Brown Skin, Springer, New Delhi, India, 2017, pp. 159–167.

[281] A.A. Anisimov, D.A. Egorov, Development of a portable device for iontophoresis, in: IEEE Conference of Russian Young Researchers in Electrical and Electronic Engineering (EIConRus), St. Petersburg and Moscow, Russia, 2020, 2020, pp. 1478–1482.

[282] F. Mandica, J. Sabattier, Device for applying a product to be distributed on the skin of a user by iontophoresis, United States Patent 10,729,899, 2020.

[283] J.H. Morriss, G. Liu, K. Hames, Iontophoresis methods, The United States Patent 10,576,277, 2020.

[284] T.G. Rukari, B.R. Alhat, Iontophoresis: an electrically assisted drug delivery system, J. Adv. Drug. Deliv. 1 (2014) 58–70.

[285] S. Arunkumar, H.N. Shivakumar, M.S. Narasimha, Effect of terpenes on transdermal iontophoretic delivery of diclofenac potassium under constant voltage, Pharm. Dev. Technol. 23 (2018) 806–814.

[286] A. Jose, S. Labala, K.M. Ninave, S.K. Gade, V.V. Venuganti, Effective skin cancer treatment by topical co-delivery of curcumin and STAT3 siRNA using cationic liposomes, AAPS PharmSciTech 19 (2018) 166–175.

[287] H. Lee, C. Song, S. Baik, D. Kim, T. Hyeon, D.H. Kim, Device-assisted transdermal drug delivery, Adv. Drug Deliv. Rev. 127 (2018) 35–45.

[288] J.M. Yoo, H.J. Park, S.W. Choi, H.O. Kim, Vitamin C-iontophoresis in melasma, Korean J. Dermatol. 39 (2001) 285–291.

[289] S. Kim, S.Y. Oh, S.H. Lee, Comparative study of glycolic acid peeling vs. vitamin C-iontophoresis in melasma, Korean J. Dermatol. 39 (2001) 1356–1363.

[290] C.H. Huh, K.I. Seo, J.Y. Park, J.G. Lim, H.C. Eun, K.C. Park, A randomized, double-blind, placebo-controlled trial of vitamin C iontophoresis in melasma, Dermatology 206 (2003) 316–320.

[291] J. Cazares Delgadillo, D. Bordeaux, Y.N. Kalia, S.S. Del Rio, J.C. Serna, Iontophoresis method of delivering vitamin c through the skin, United States Patent 16/648,038, 2020.

[292] I. Kurokawa, Comprehensive sequential successful therapy comprising chemical peeling, iontophoresis and topical vitamin C for postinflammatory hyperpigmentation in acne vulgaris, J. Cosmet. Dermatol. Sci. Appl. 9 (2020) 99–103.

[293] I. Kurokawa, Non-surgical treatment with chemical peeling and subsequent vitamin C iontophoresis for rolling scars in acne vulgaris, J. Cosmet. Dermatol. Sci. Appl. 10 (2020) 104–106.

[294] R.M. Sobhi, A.M. Sobhi, A single-blinded comparative study between the use of glycolic acid 70% peel and the use of topical nanosome vitamin C iontophoresis in the treatment of melasma, J. Cosmet. Dermatol. 11 (2012) 65–71.

[295] M.B. Taylor, J.S. Yanaki, D.O. Draper, J.C. Shurtz, M. Coglianese, Successful short-term and longterm treatment of melasma and postinflammatory hyperpigmentation using vitamin C with a full-face iontophoresis mask and a mandelic/malic acid skin care regimen, J. Drugs Dermatol. 12 (2013) 45–50.

[296] Y.K. Choi, Y.K. Rho, K.H. Yoo, Y.Y. Lim, K. Li, B.J. Kim, S.J. Seo, M.N. Kim, C.K. Hong, D.S. Kim, Effects of vitamin C vs. multivitamin on melanogenesis: comparative study in vitro and in vivo, Int. J. Dermatol. 49 (2010) 218–226.

[297] R.J. DeMartini, Epidermal iontophoresis device, United States Patent 4,997,418, 1991.

[298] G.M. Gelfuso, T. Gratieri, M.B. Delgado-Charro, R.H. Guy, R.F. Vianna Lopez, Iontophoresis-targeted, follicular delivery of minoxidil sulfate for the treatment of alopecia, J. Pharm. Sci. 102 (2013) 1488–1494.

[299] S. Narasimha Murthy, D.E. Wiskirchen, B.C. Paul, Iontophoretic drug delivery across human nail, J. Pharm. Sci. 96 (2007) 305–311.

[300] P.P. Shanbhag, U. Jani, Drug delivery through nails: present and future, New Horiz. Transl. Med. 3 (2017) 252–263.

[301] A. Kushwaha, H.N. Shivakumar, S.N. Murthy, Iontophoresis for drug delivery into the nail apparatus: exploring hyponychium as the site of delivery, Drug Dev. Ind. Pharm. 42 (2016) 1678–1682.

[302] R. Aggarwal, M. Targhotra, B. Kumar, P.K. Sahoo, M.K. Chauhan, Treatment and management strategies of onychomycosis, J. Mycol. Med. 30 (2020) 1–15.

[303] S.N. Murthy, D.C. Waddell, H.N. Shivakumar, A. Balaji, C.P. Bowers, Iontophoretic permselective property of human nail, J. Dermatol. Sci. 46 (2007) 150–152.

[304] T. Gratieri, G.M. Gelfuso, Overcoming hurdles in iontophoretic drug delivery: is skin the only barrier? Ther. Deliv. 5 (2014) 493–496.

[305] M.B. Brown, R.H. Khengar, R.B. Turner, B. Forbes, M.J. Traynor, C.R. Evans, S.A. Jones, Overcoming the nail barrier: a systematic investigation of ungual chemical penetration enhancement, Int. J. Pharm. 370 (2009) 61–67.

[306] P. Chouhan, T.R. Saini, Hydroxypropyl-β-cyclodextrin: a novel transungual permeation enhancer for development of topical drug delivery system for onychomycosis, J. Drug Deliv. 2014 (2014) 1–7.

[307] E.A. Bseiso, M. Nasr, O.A. Sammour, N.A. Abd El Gawad, Novel nail penetration enhancer containing vesicles "nPEVs" for treatment of onychomycosis, Drug Deliv. 23 (2016) 2813–2819.

[308] E. Cutrín-Gómez, S. Anguiano-Igea, M.B. Delgado-Charro, J.L. Gómez-Amoza, F.J. Otero-Espinar, Effect of penetration enhancers on drug nail permeability from cyclodextrin/poloxamer-soluble polypseudorotaxane-based nail lacquers, Pharmaceutics 10 (2018) 273–291.

[309] N. Saki, S. Hosseinpoor, A. Heiran, A. Mohammadi, M. Zeraatpishe, Comparing the efficacy of triamcinolone acetonide iontophoresis versus topical calcipotriol/betamethasone dipropionate in treating nail psoriasis: a bilateral controlled clinical trial, Dermatol. Res. Pract. 2018 (2018) 1–7.

[310] M. Bok, Z.J. Zhao, S. Jeon, J.H. Jeong, E. Lim, Ultrasonically and iontophoretically enhanced drug-delivery system based on dissolving microneedle patches, Sci. Rep. 10 (2020) 1–10.

[311] C. Curdy, Y.N. Kalia, R.H. Guy, Non-invasive assessment of the effects of iontophoresis on human skin in-vivo, J. Pharm. Pharmacol. 53 (2001) 769–777.

[312] J. Singh, M. Gross, B. Sage, H.T. Davis, H.I. Maibach, Regional variations in skin barrier function and cutaneous irritation due to iontophoresis in human subjects, Food Chem. Toxicol. 39 (2001) 1079–1086.

[313] J.J. Escobar-Chavez, D. Bonilla-Martínez, M.A. Villegas-González, I.M. Rodríguez-Cruz, C.L. Domínguez-Delgado, The use of sonophoresis in the administration of drugs throughout the skin, J. Pharm. Pharm. Sci. 12 (2009) 88–115.

[314] B.E. Polat, D. Hart, R. Langer, D. Blankschtein, Ultrasound-mediated transdermal drug delivery: mechanisms, scope, and emerging trends, J. Control. Release 152 (2011) 330–348.

[315] A. Azagury, L. Khoury, G. Enden, J. Kost, Ultrasound mediated transdermal drug delivery, Adv. Drug Deliv. Rev. 72 (2014) 127–143.

[316] S. Mitragotri, Sonophoresis: ultrasound-mediated transdermal drug delivery, in: N. Dragicevic, H.I. Maibach (Eds.), Percutaneous Penetration Enhancers Physical Methods in Penetration Enhancement, Springer, Berlin, Germany, 2017, pp. 3–14.

[317] S. Daftardar, R. Neupane, S.H. Boddu, J. Renukuntla, A.K. Tiwari, Advances in ultrasound mediated transdermal drug delivery, Curr. Pharm. Des. 25 (2019) 413–423.

[318] P. Santoianni, M. Nino, G. Calabro, Intradermal drug delivery by low frequency sonophoresis (25KHz), Dermatol. Online J. 10 (2004) 24.

[319] S.H. Lee, C.M. Wang, C.J. Chung, H.S. Hong, Sonophoresis with 20% L-ascorbic acid and 2% kojic acid gel for the melasma patient, Chin. J. Dermatol. 19 (2001) 275–281.

[320] H.F. Liu, Z.Z. Liu, Sonophoresis on levorotatory vitamin C combined with the whitening essence of arbutin B3 in the treatment of chloasma clinical observation of 46 cases, Chin. J. Aesthet. Med. 2013 (2013) 20.

[321] L. Hsin-Ti, L. Wen-Sheng, W. Yi-Chia, L. Ya-Wei, Z.H. Wen, W.H. David, L. Su-Shin, The effect in topical use of LycogenTM via sonophoresis for anti-aging on facial skin, Curr. Pharm. Biotechnol. 16 (2015) 1063–1069.

[322] M. Zasada, Z. Drożdż, A. Erkiert-Polguj, E. Budzisz, A blinded study assessment of the efficacy of an original formula with retinol in combination with sonophoresis, Dermatol. Ther. 33 (2020) e13163.

[323] K.A. Laudati, Rose quartz and/or jade multi-roller applicator and container apparatus, The United States Patent 16/258,236, 2020.

[324] W. Yi, Silent jade massage cosmetic roller that can be freely assembled, The United States Patent 16/285,129, 2020.

Cumulative index

Bold numerals refer to volume numbers.